Y0-BXV-855

4.13.77

NEW AND EASY QUILTING

Sibby Taylor

Photographs by Bob Mehesy

South Brunswick and New York: A.S. Barnes and Company
London: Thomas Yoseloff Ltd

A. S. Barnes and Co., Inc.
Cranbury, New Jersey 08512

Thomas Yoseloff Ltd
Magdalen House
136-148 Tooley Street
London SE1 2TT, England

Library of Congress Cataloging in Publication Data

Taylor, Sibby, 1943–
New and easy quilting.

Bibliography: p.
Includes index.
1. Quilting. I. Title.
TT835.T34 1977 746.9'7 76-27492
ISBN 0-498-02001-0

PRINTED IN THE UNITED STATES OF AMERICA

To Mary Beth, who started the whole thing.

To my parents, Frank and Ruby Owens for *not* saying all the things they wanted to when I told them I was quitting my job to make a living teaching quilting.

To the members of my first class who stuck it out even when my daughter's pet snake got loose in class.

And to all my other students, who have taught me so much.

Contents

Preface, 9
1 Historical Background, 11
2 Piecing, 14
3 Appliqué, 20
4 Quilting, 28
5 White-on-White Quilting, 35
6 Block Method Construction, 37
7 Binding, 40
8 Tabletop Quilting, 44
9 Quilt Planning, 46
10 Hawaiian Quilting, 49
11 Novelties, 51
12 Restorations, 57
Completed Designs, 58
Patterns, 64
Bibliography, 110
Index, 111

Preface

My students and customers almost always ask me where I learned to quilt. They assume that because I am from the South that I must have been taught by my grandmother or mother. The truth is I never saw anyone quilt.

As a child I slept in a huge antique bed covered with a pieced quilt, and when I had the bed shipped to New Jersey, my mother included the quilt. It was in such bad condition that I decided to make one for the bed to replace it. And that's how I learned to quilt.

My first quilt taught me many things not to do—it took me nearly a year to make. My second taught me still more, and every quilt since has increased my knowledge.

At the time I began quilting there were no books to help me, and although I study each new one published, I still haven't found one I feel explains all the quick tricks and easy ways to do things. That's why I'm writing this book—to help you begin quilting the easy way, to help you avoid the mistakes I made, and to correct the ones you will make.

Very comforting is the superstition among long-ago quilters that it was an offense against God to make a perfect quilt. They often reversed a block on purpose to achieve the nonperfect result.

It is also comforting to know that some quilts seen in museums were made over many years by women who had nothing else to do. Strive to match their effects, not their hours of work.

Above all, I hope you'll enjoy quilting as much as I do.

1

Historical Background

It is thought that quilting originated in China centuries ago and slowly worked its way westward until the time of the Crusades, when it had reached Jerusalem. The European knights, in the aftermath of battle, discovered that Moslem warriors wore quilted undergarments to protect their skin from being chafed by armor.

These European knights took the idea of quilting with them when they returned home, and their wives adapted it to everything from window curtains to skirts and jackets. Europe needed warmth and quilting provided it. The wealthy wore clothes of silk padded with wool and quilted, sometimes with gold and silver thread. Peasants used homespun, stuffed with straw and quilted. Bedcoverings were always one piece of fabric, stuffed, lined with another piece of fabric, and quilted.

After the American colonies were founded, one great shortage was cloth. In the beginning all fabric was imported at tremendous cost, or woven by hand in the home; therefore, every scrap was precious. Colonial housewives cut down their worn clothing and that of their husbands to make clothes for their children. When those clothes were completely worn out, they salvaged whatever scraps of fabric were still good and sewed them together in random patterns and quilted them to make warm bedcoverings. These first quilts were stuffed with wool (if sheep were available) or cotton (if made in the South).

Soon these same housewives discovered it was easier and not too wasteful of fabric to cut scraps into squares, rectangles, and triangles, and the first patchwork patterns were born.

These pieced patterns were named for familiar objects in the home, occupations of the day, objects in nature, and biblical subjects. Their names remind us of the great history of our country—Bearpaw, Log Cabin, Tomahawk, Washington Pavement, Lincoln's Platform, World's Fair.

The same quilt block was often known by different names depending upon the area in which they were made. The most famous example is Bearpaw, also known as Hand of Friendship (in Philadelphia) and Ducks-Foot-in-the-Mud (on Long Island). Some names were so popular that numerous blocks were given the same name. For example, there are many different blocks known as Rose of Sharon.

The next development in American quilting occurred in the 1800s with the introduction of fabric manufacturing on a commercial basis. With cheaper, more widely available cloth, American quilters turned to appliqué for their finest quilts.

Crazy block.

Appliqué is very wasteful of fabric since it consists of cloth laid on cloth, sometimes several layers, which is then hemmed down. Appliqué was usually reserved for bridal, friendship, or presentation quilts, which were intended mostly for "show" and never used. This must be the reason more antique appliqué quilts survive than pieced ones.

Quilting fell out of fashion with the introduction of inexpensive, machine-made quilts and bedspreads, but had a brief revival during the Depression of the 1930s. With the approach of the Bicentennial, quilting experienced a renaissance.

Bearpaw block.

2 Piecing

Piecing consists of sewing together various-shaped pieces of fabric to form blocks, usually square or rectangular. The simplest pieced block is the Nine-Patch, basis of many more complicated patterns. In this pattern, squares are first sewed together in rows, then the rows sewed together to form a square block. Seams of ¼ inch are always used. Stitching can be by machine, 8 to 10 stitches to the inch, or by hand, using a fine running stitch or a backstitch. Polyester thread is preferred when sewing by machine, cotton when sewing by hand. After each seam is stitched, it should be trimmed even. Pressing seams open or to one side is not necessary. When seaming rows together, match and pin seam lines, easing where necessary.

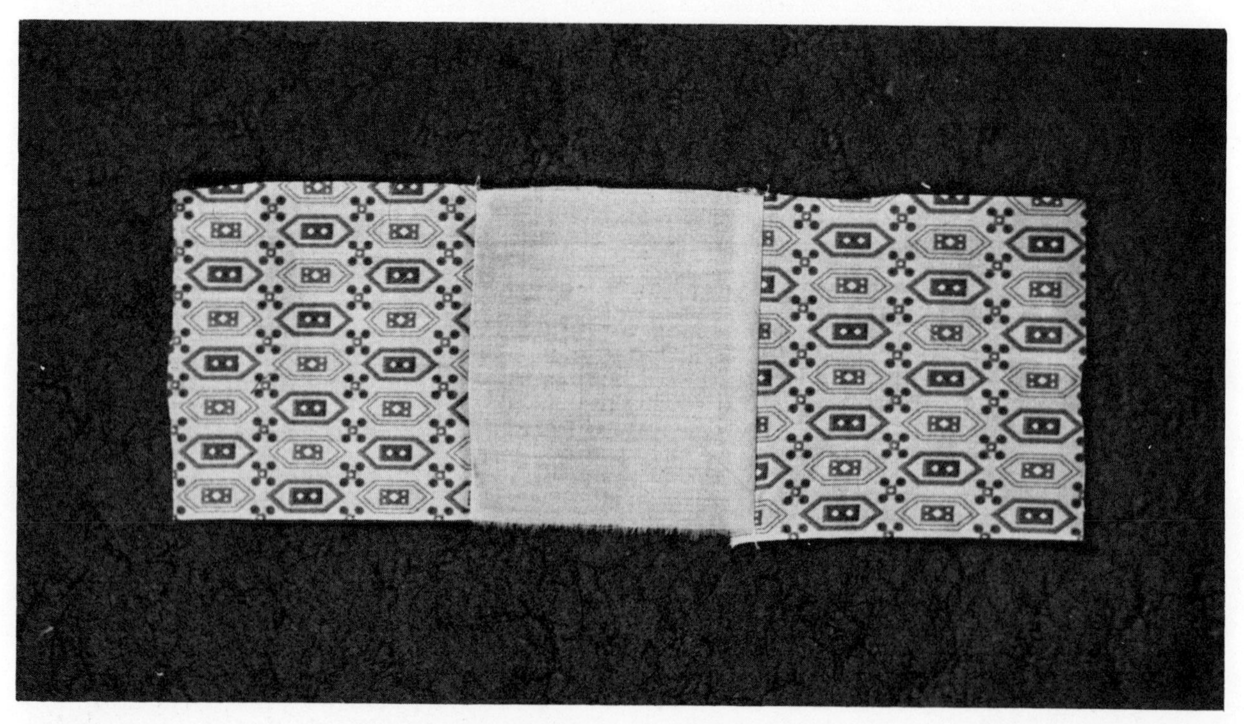

Three squares are sewed together into a row.

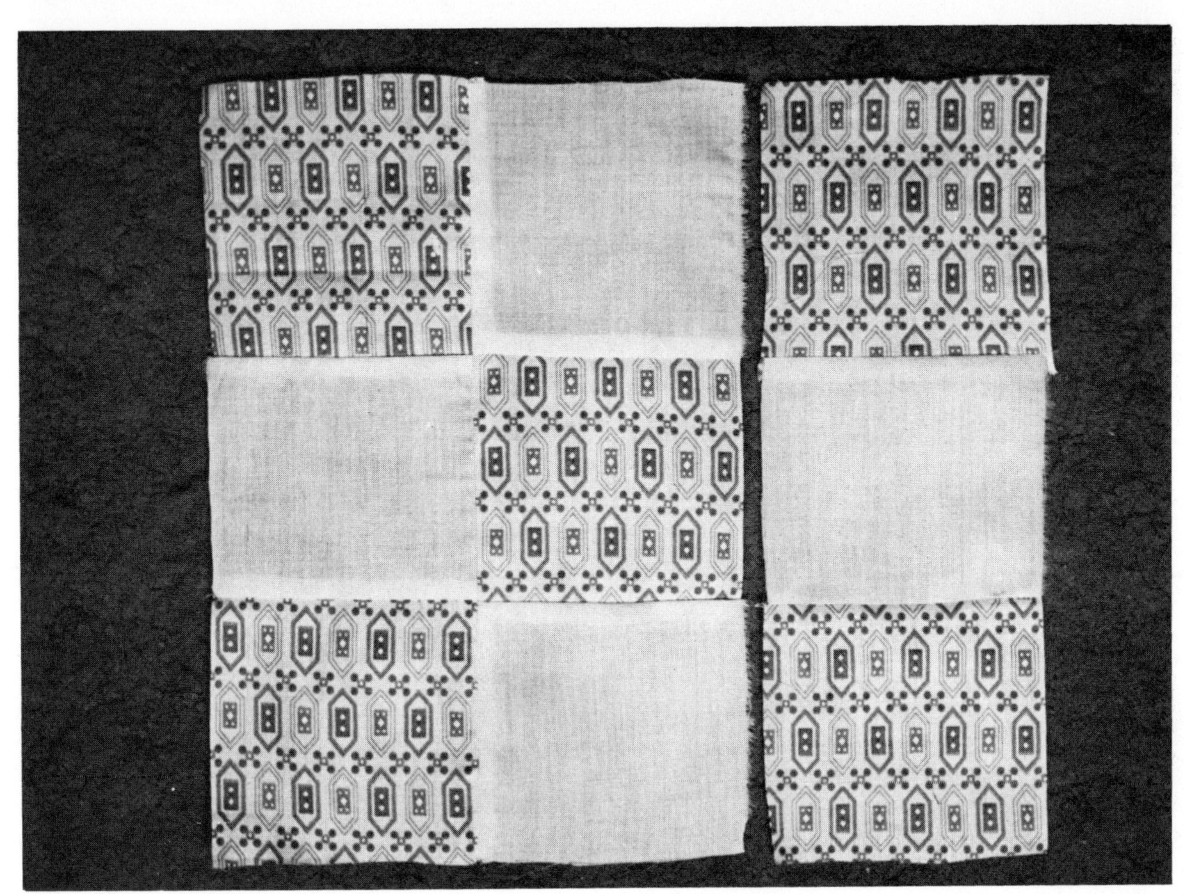

When all three rows are completed, rows 1 and 2 are sewed together.

Completed Nine-Patch block with all rows sewed together.

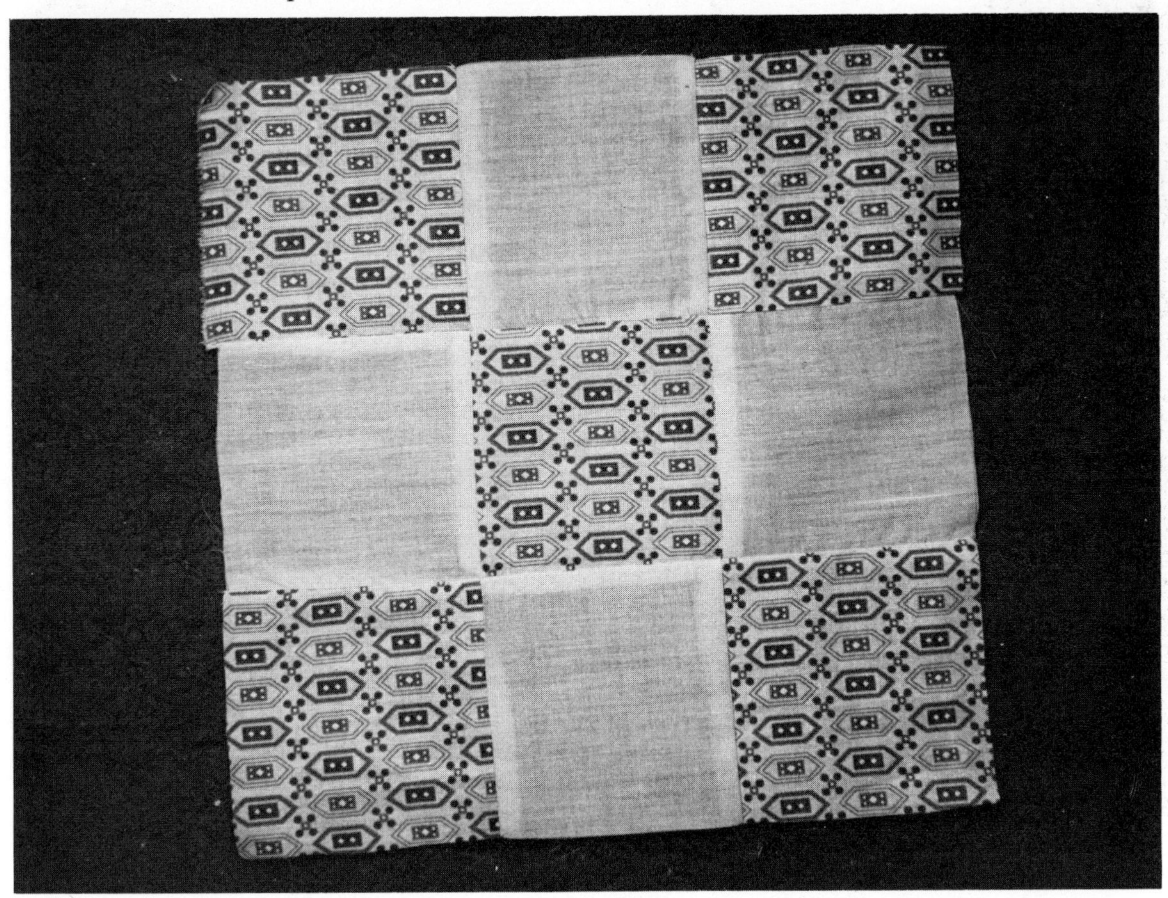

A light and a dark triangle sewed together forming a square, and the completed first row.

Completed Shoo-Fly block.

A slightly more difficult version of the Nine-Patch is Shoo-Fly. It is assembled identically to the Nine-Patch, except that a first step of sewing the light and dark triangles together along the long side (the hypotenuse) is necessary.

Even more complicated patterns such as Lincoln's Platform can be put together row method.

Separate sections of Lincoln's Platform block sewed together and ready for assembly.

Completed Lincoln's Platform block. *Courtesy Marcia Forrest.*

Building out from a center piece is used in such patterns as Log Cabin and Modern Geometric. The construction of Log Cabin is as follows:

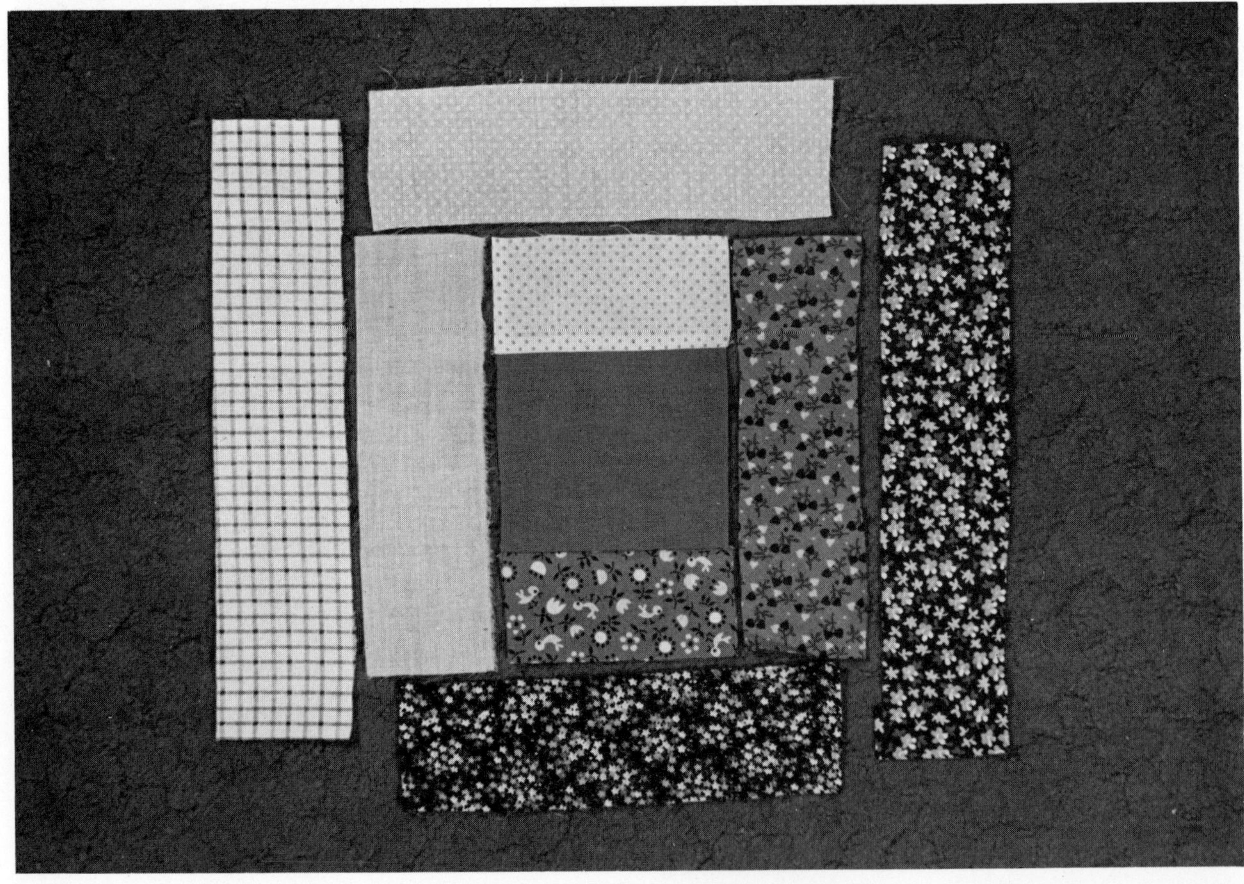

Expanded view of Log Cabin block.

Modern Geometric is also built out from a center.

Expanded view of Modern Geometric block.

Fabrics should be lightweight cottons or polyester and cotton blends. Always check to see that fabrics are color-fast and nonshrink. If in doubt, you should test colors and preshrink fabric yourself.

Pattern pieces should be cut out of lightweight cardboard, cereal boxes for example, or sandpaper. If sandpaper is used, place gritty side down on fabric when tracing pattern piece onto fabric. Draw around pattern piece with pencil, chalk, or a sliver of soap or use laundry marker, depending upon the color of the fabric. Do not use ball-point pen, since it tends to smear and will run when washed. Never try to fold fabric and cut out several pieces at one time. Accuracy is extremely important in piecing. If edges are not true, the block will not go together correctly.

All pieced patterns in this book have ¼-inch seam allowance included.

3 Appliqué

Modern quilters seem to prefer appliqué to piecing for their finest work, probably because appliqué offers more scope for originality in design and execution. After an appliqué pattern has been selected, pattern pieces are prepared of cardboard or sandpaper, the pattern is traced onto fabric, and the pieces cut out. Appliqué pieces can be cut in two different ways. In either case, they should be cut approximately ¼ inch larger than planned finished size. Appliqué stitching should *always* be done by hand.

The first method of cutting out appliqué pieces consists of drawing the pattern actual finished size. These are placed on fabric so that at least ½ inch is free on all sides. Pencil or laundry marker is used to draw around each piece, then they are cut out leaving ¼-inch allowance on all sides.

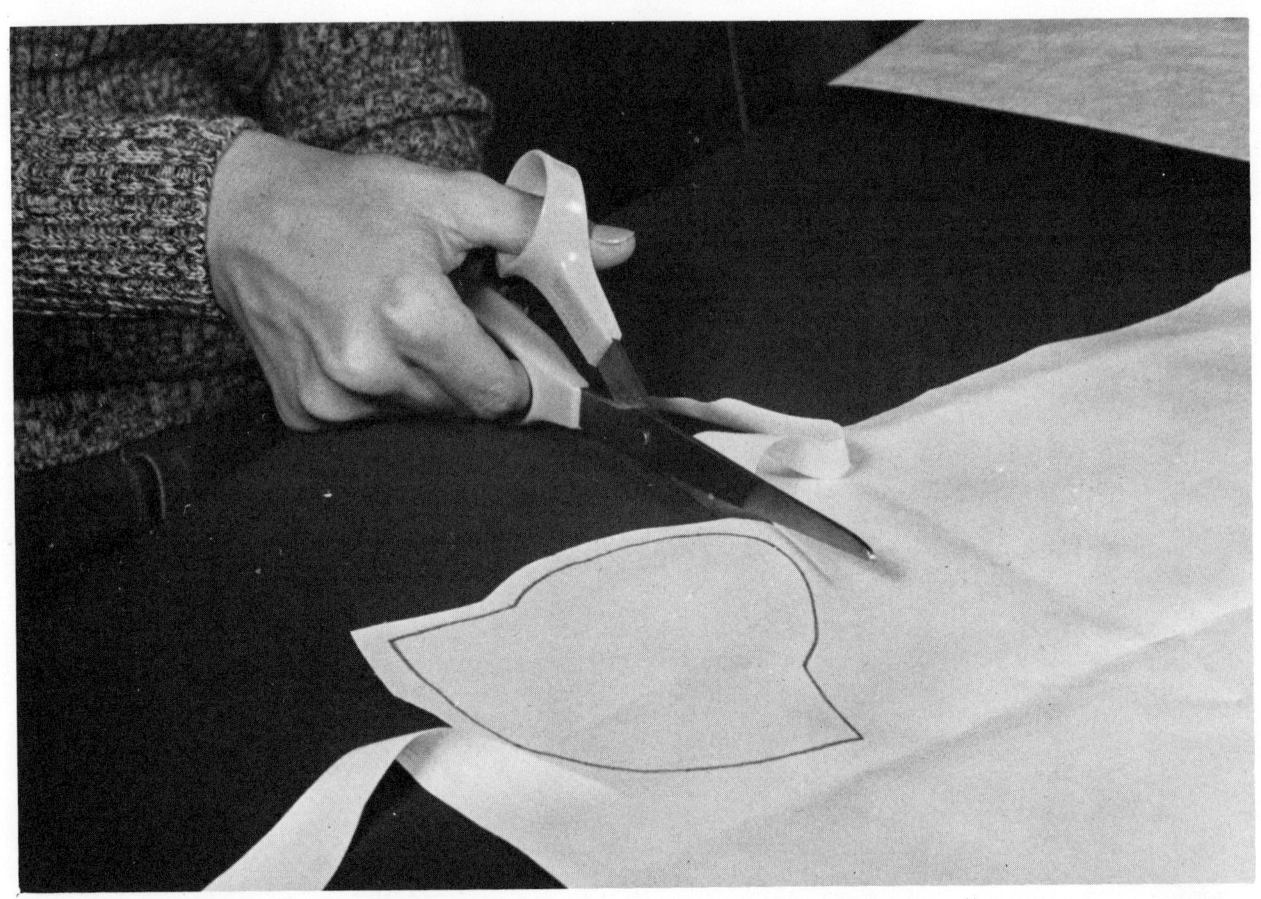

Method No. 1 of cutting out appliqué pieces.

The second method of cutting out appliqué pieces is to draw pattern pieces actual finished size, then carefully add ¼ inch all around. Pattern is placed on fabric, drawn around, then cut out along pencil line.

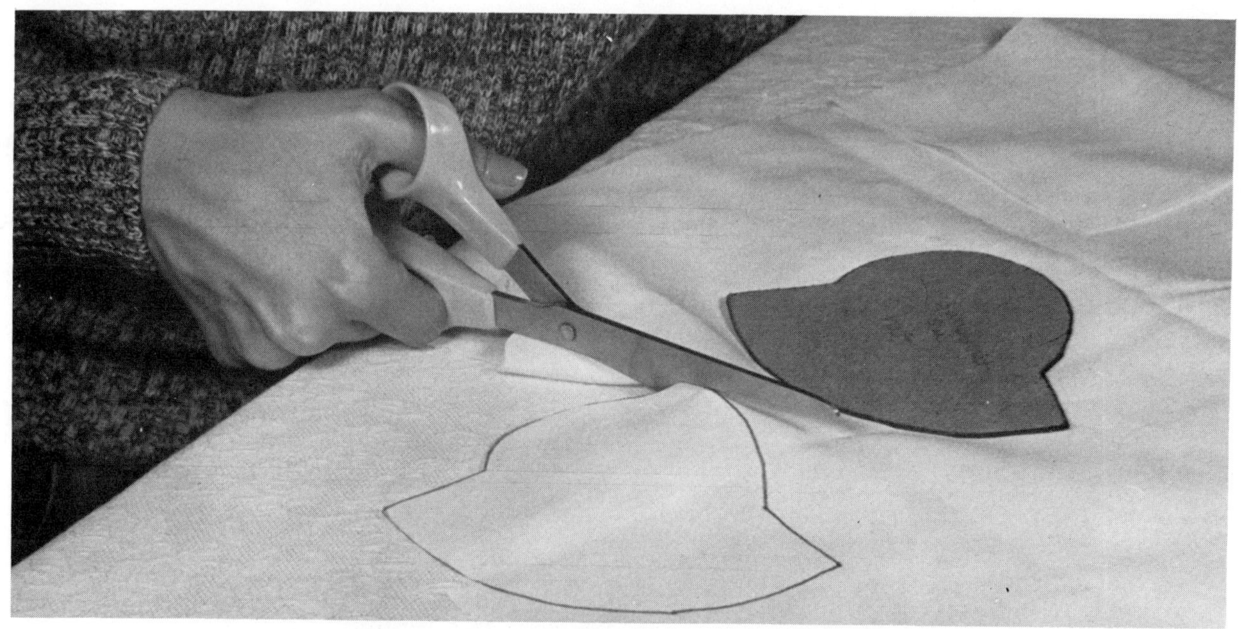

Method No. 2 of cutting out appliqué pieces.

When all pieces have been cut out, pieces are pinned securely to background block and sewed to it. Appliqué can be done in five different ways, the easiest being use of a fine running stitch near the turned under edge.

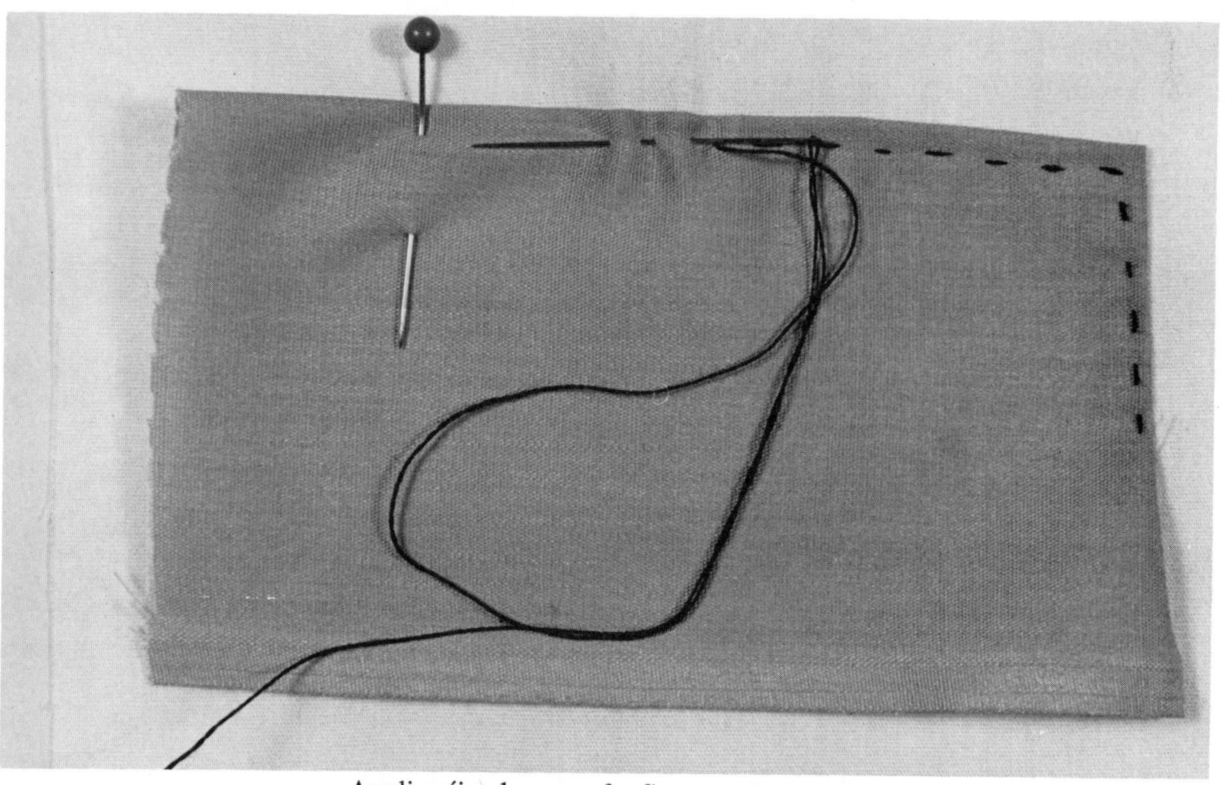

Appliquéing by use of a fine running stitch.

Slightly more difficult is use of a fine hemming stitch, the same used to hem handkerchiefs and scarves. I grew up calling this the whip stitch.

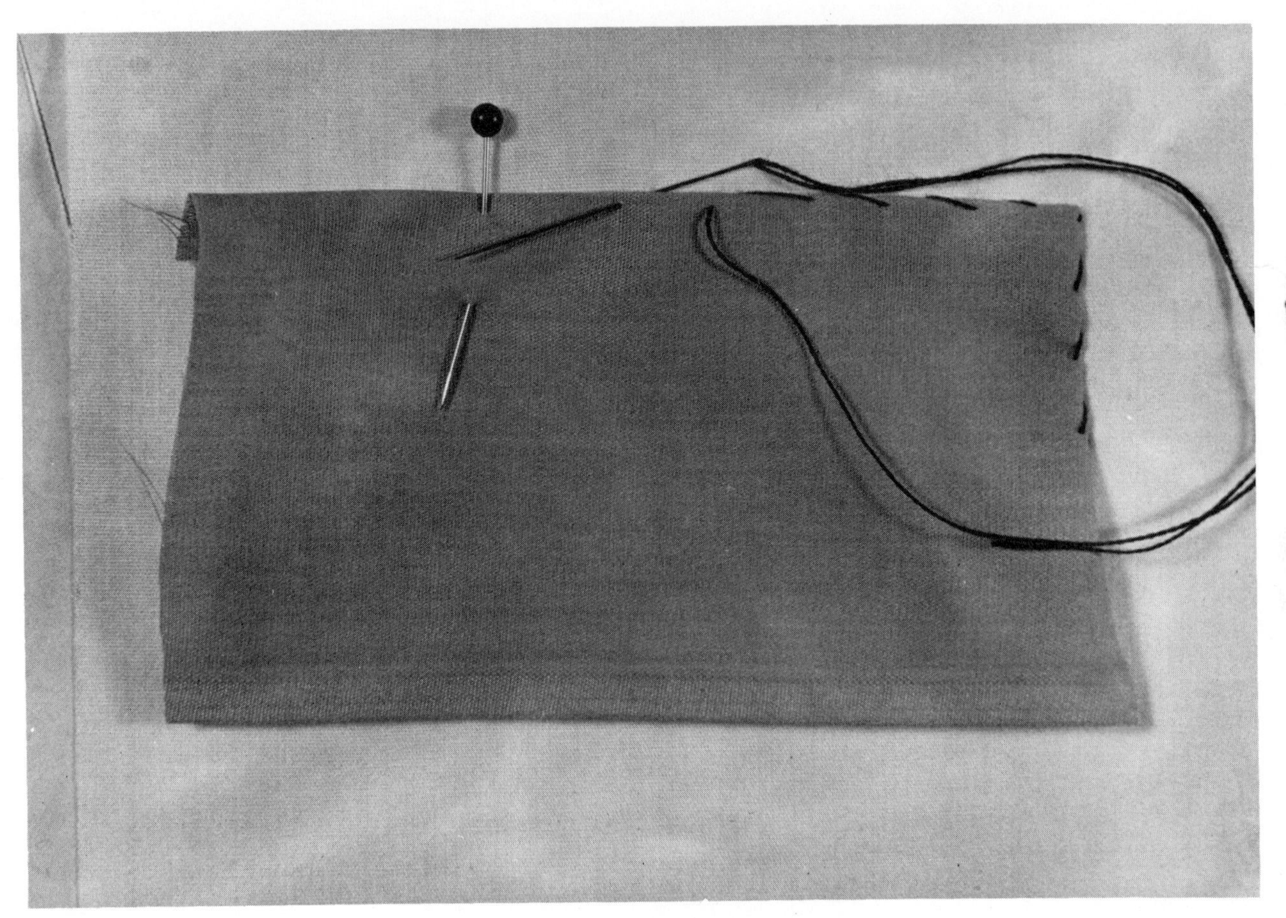

Appliquéing by use of a whip stitch.

In the preceding two methods, thread color should always match the appliqué piece, since thread is visible.

Quite difficult unless practiced is the blind stitch. Still, it is worth the effort, since the finest appliqué is always done with the blind stitch. One of my elderly pupils told me that she learned to blind stitch in the eighth grade and passed her final test in it when her teacher had her blind stitch a white appliqué with black thread! Any shade of thread may be used to do the blind stitch, since the thread is never seen. You may wish to use pale thread with pale appliqué pieces and dark thread with dark ones.

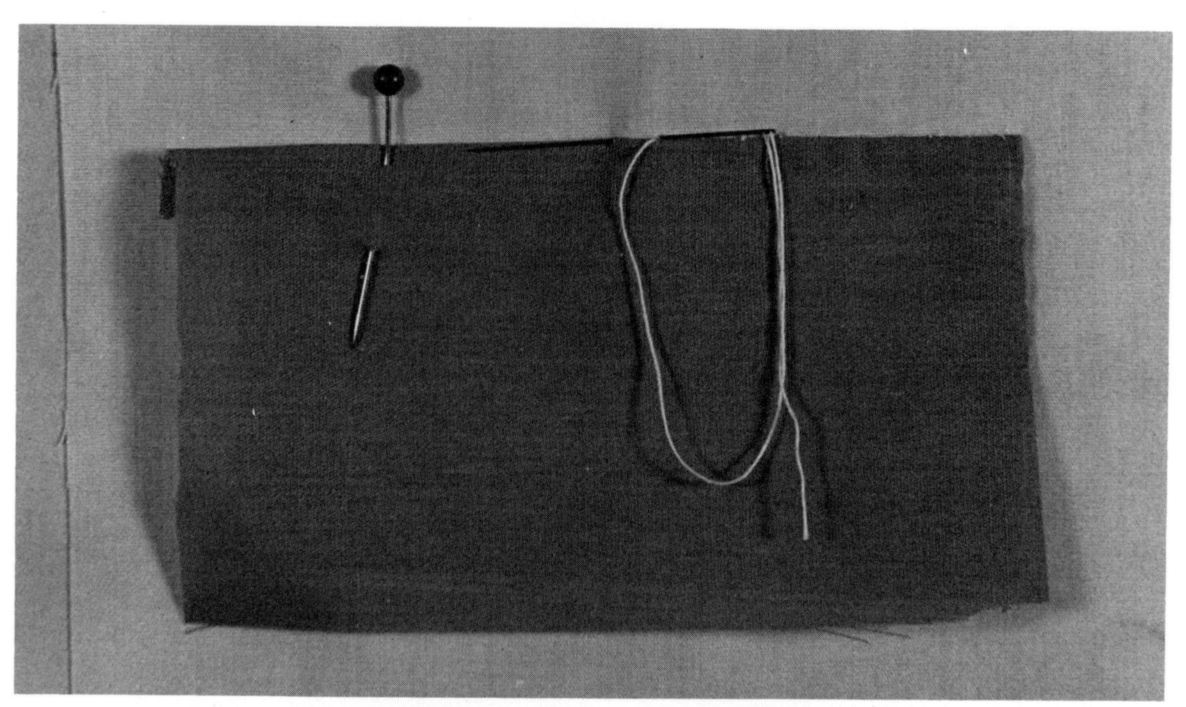

Appliquéing by use of the blind stitch.

The final two methods of appliqué are different in that appliqué pieces are cut out actual finished size and the raw edges are covered with embroidery stitches. Two or three strands of embroidery floss are used, matching or contrasting with the appliqué piece.

Buttonhole stitch, sometimes called blanket stitch, is done with the individual stitches set close together and made as small as possible.

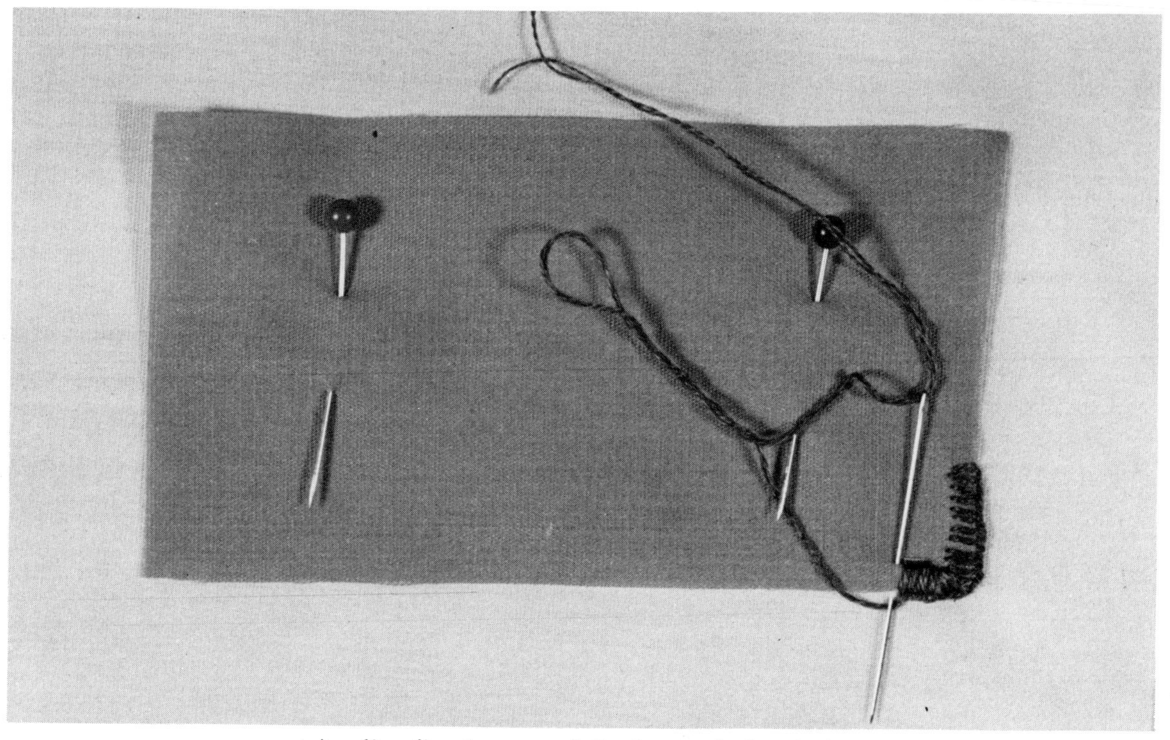

Appliquéing by use of the buttonhole stitch.

The last method of applying pieces is use of the satin stitch.

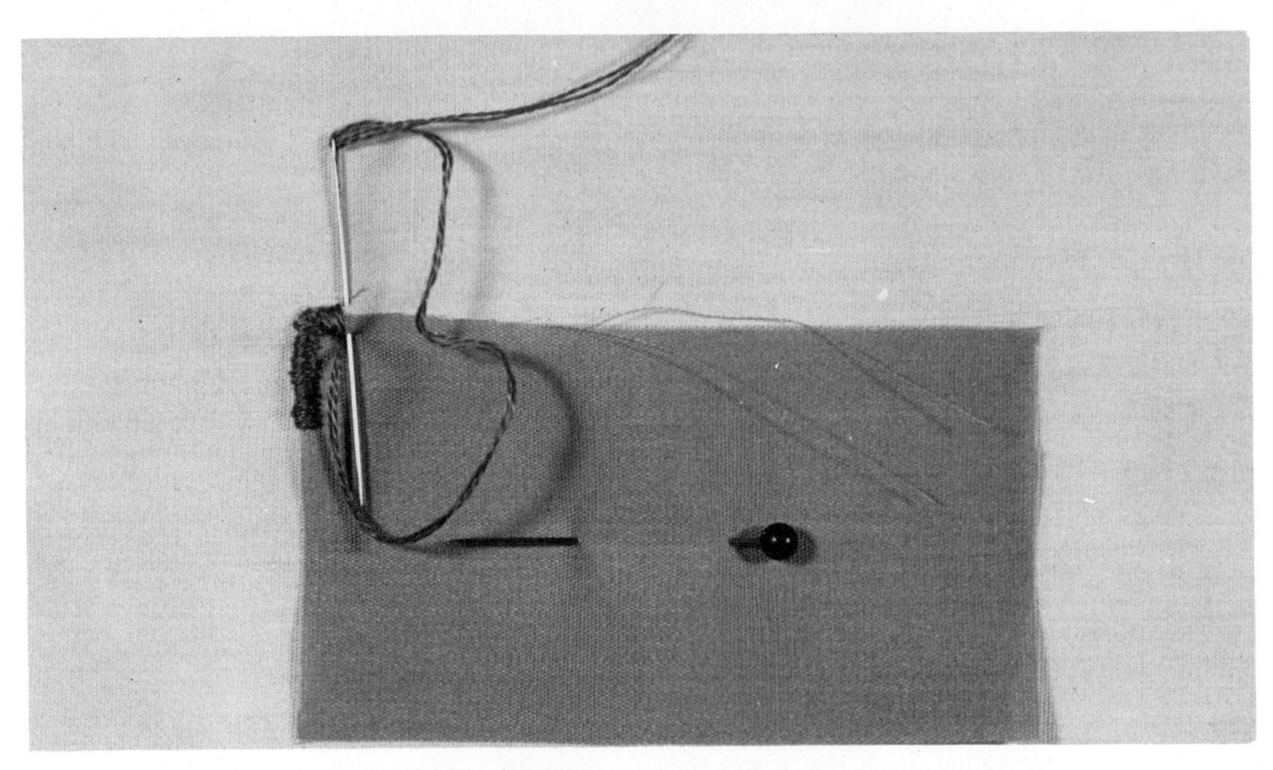

Appliquéing by use of the satin stitch.

Corners and curves of appliqué pieces are not difficult to do; however, there are a few useful hints for points and indentations. Points should be tapered as much as possible during the turning-under process as shown.

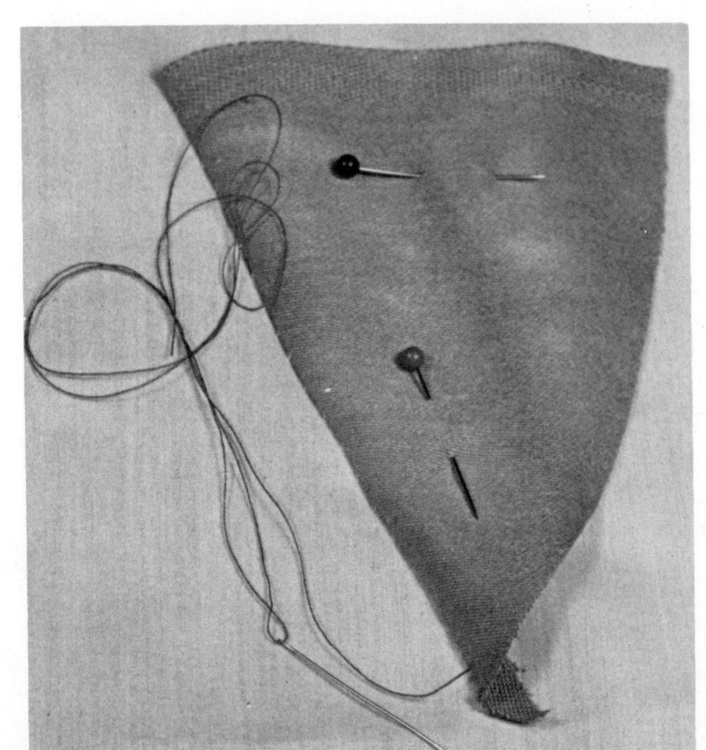

Blind stitching to within ¼" of the point.

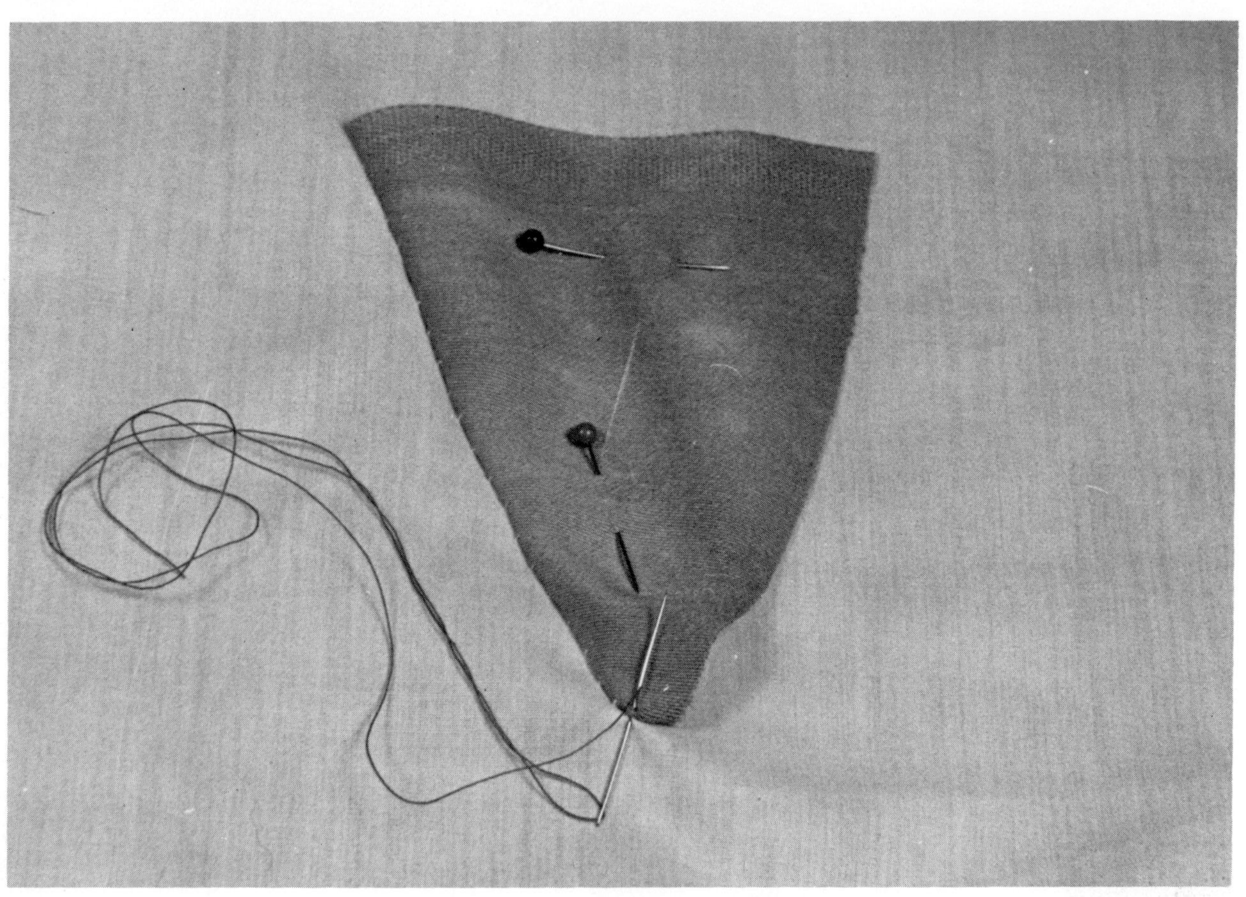

Tacking the tip of the point.

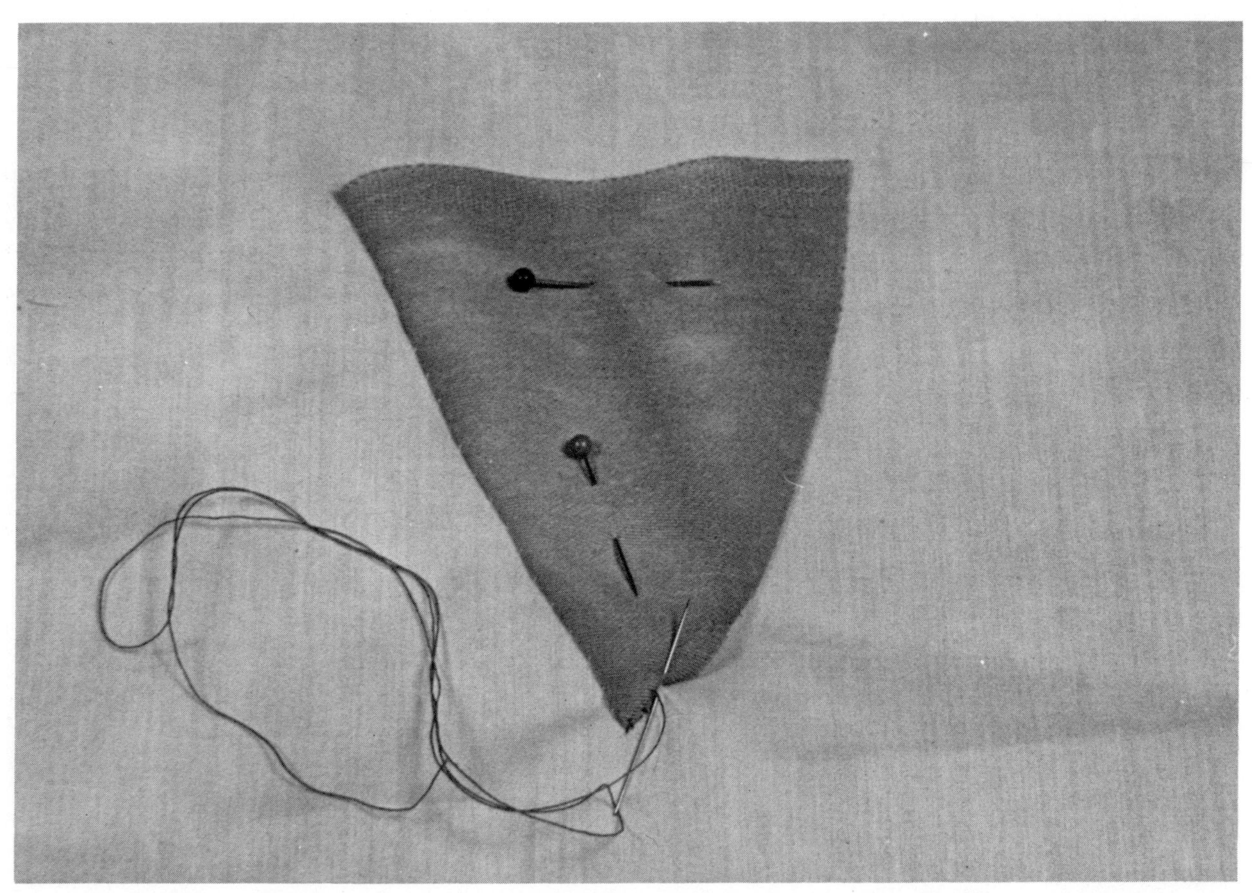

Blind stitch continues using the needle to force under the ¼".

Indentations should first be slashed and carefully tacked with two or three stitches before continuing the appliqué.

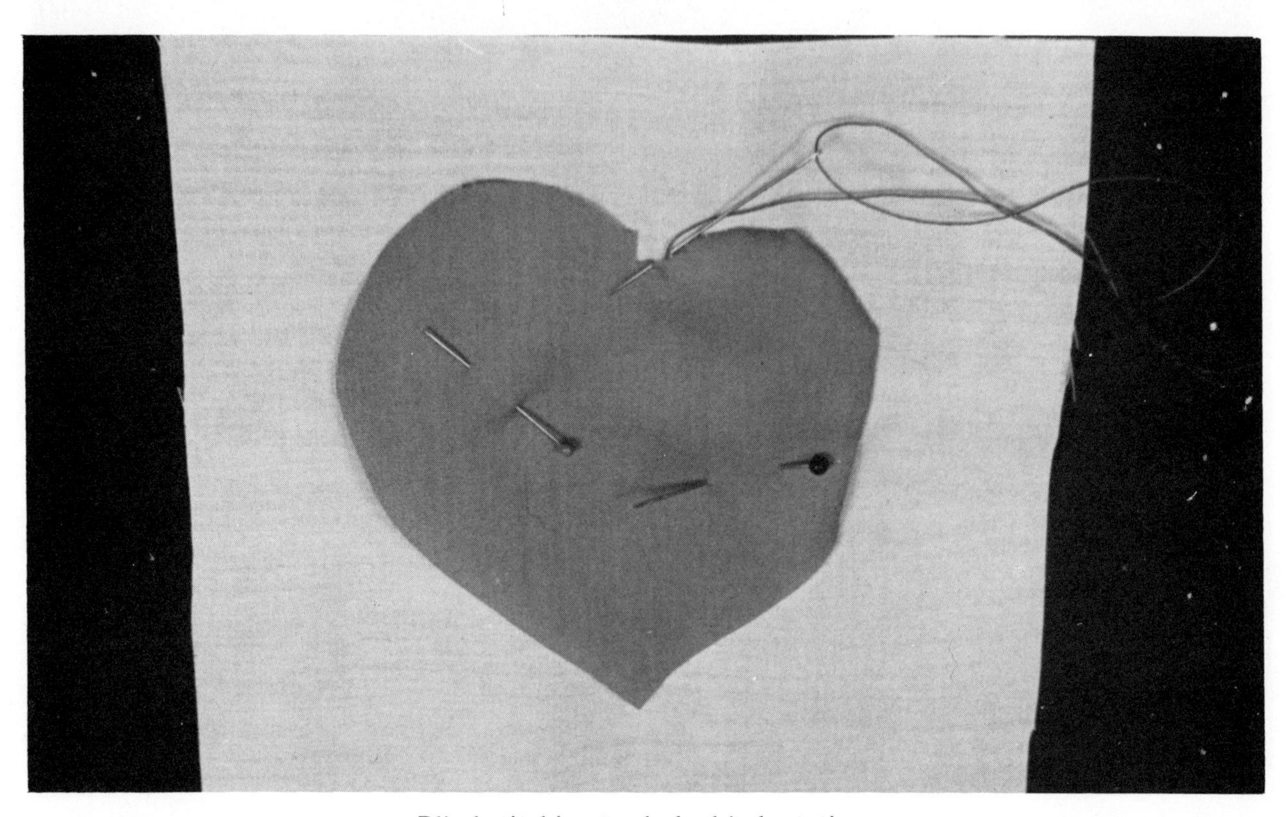

Blind stitching to slashed indentation.

Tacking indentation with several small stitches.

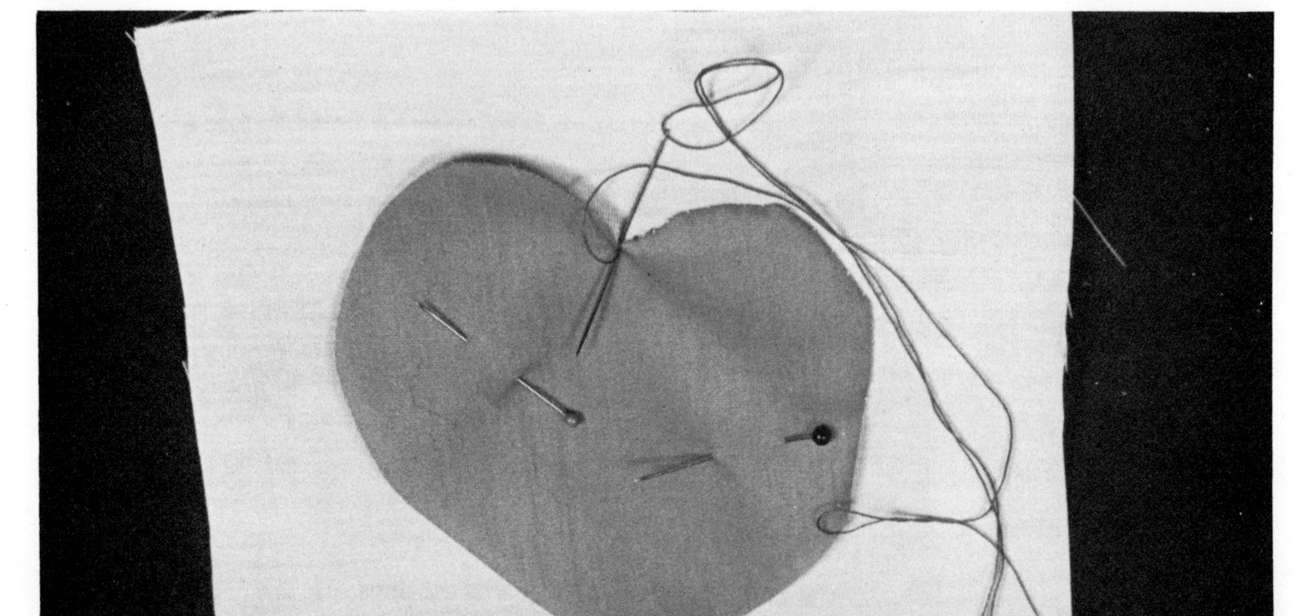

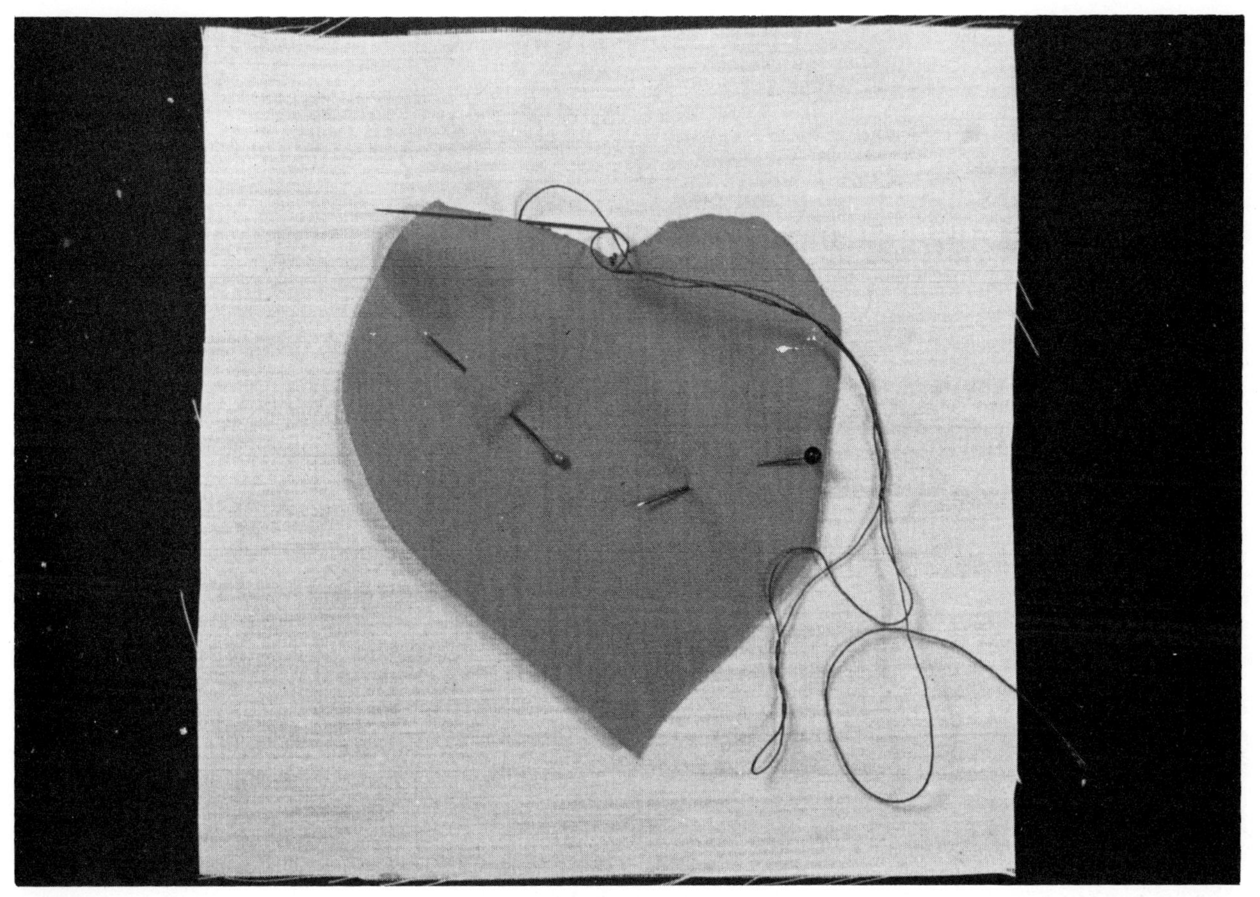

Continuing blind stitching.

Stay stitching along pencil line of Method 1 appliqué or ¼ inch inside cut edge of Method 2 appliqué is useful both as a turn-under guide and to prevent fraying when slashing must be done.

After sufficient practice, the ¼-inch turn-under can be done "by eye" and at the same time as appliquéing. If this is too difficult at first, the ¼-inch edge of each piece may be pinned or basted under before actual appliquéing is done.

Original designs for appliqué are not difficult to create. Theme quilts such as the Album quilt illustrated in this book or the Noah's Ark wall hanging are good, simple ideas. Try a four-block wall hanging called The Seasons with each block the same scene at different times of year. Fairy tales and nursery rhymes are inspiration for children's quilts. If you cannot draw or find someone to draw for you, copy patterns line for line from primary coloring books.

Simple embroidery such as satin stitch and outline stitch, and french knots, are useful for adding details.

All appliqué patterns in this book are actual size and look well on a 15-inch background block.

4 Quilting

Before quilting can be done, three layers must be pinned or basted together. These three layers are: (1) completed pieced or appliquéd block, (2) quilt batt, (3) backing fabric cut slightly larger (about ½ inch on all sides) than completed block.

Quilt batt can be flannel, cotton batt, or polyester batt. Having worked with all three, I recommend using polyester batt only. Flannel is heavy, somewhat difficult to sew through, and takes much longer than the other to dry. Cotton batt gives the completed quilt an antique appearance, but must be quilted closely, since cotton batt shifts and lumps when washed or cleaned. On the other hand, polyester batting is lightweight but warm, comes in several thicknesses, never lumps or shifts, and machine washes, machine dries, and dry cleans perfectly. It does not have to be quilted closely, thereby cutting your work in half.

The layers are laid in the right order on top of each other on a table top, then pinned together at each corner. Basting should be done with huge stitches completely through all layers as follows:

Block basted ready for quilting.

As shown in the illustration, basting begins in the center and continues to each edge. Then a line of basting stitches is placed along the outer edge. Basting is removed when quilting is completed. White thread is always used for basting, since colored thread may mark fabric when removed. After you have become more adept at quilting, basting is no longer necessary. Pinning at each corner and the center will probably be sufficient. Polyester quilt batt can be pieced by simply placing it edge to edge on the backing fabric and basting all layers together very carefully.

The quilting stitch is a small running stitch that goes completely through all three layers. Right-handed persons should guide the needle with their right hand, while keeping the fingers of their left hand underneath the block. These fingers will feel when the needle has gone through all the layers to the back and may be turned up again. It is useful to wear a thimble, if possible. There appears to be no way to protect the "quilting finger" on your left hand–the one constantly pricked by the needle as it passes through the layers. In time a callous will form to protect that finger.

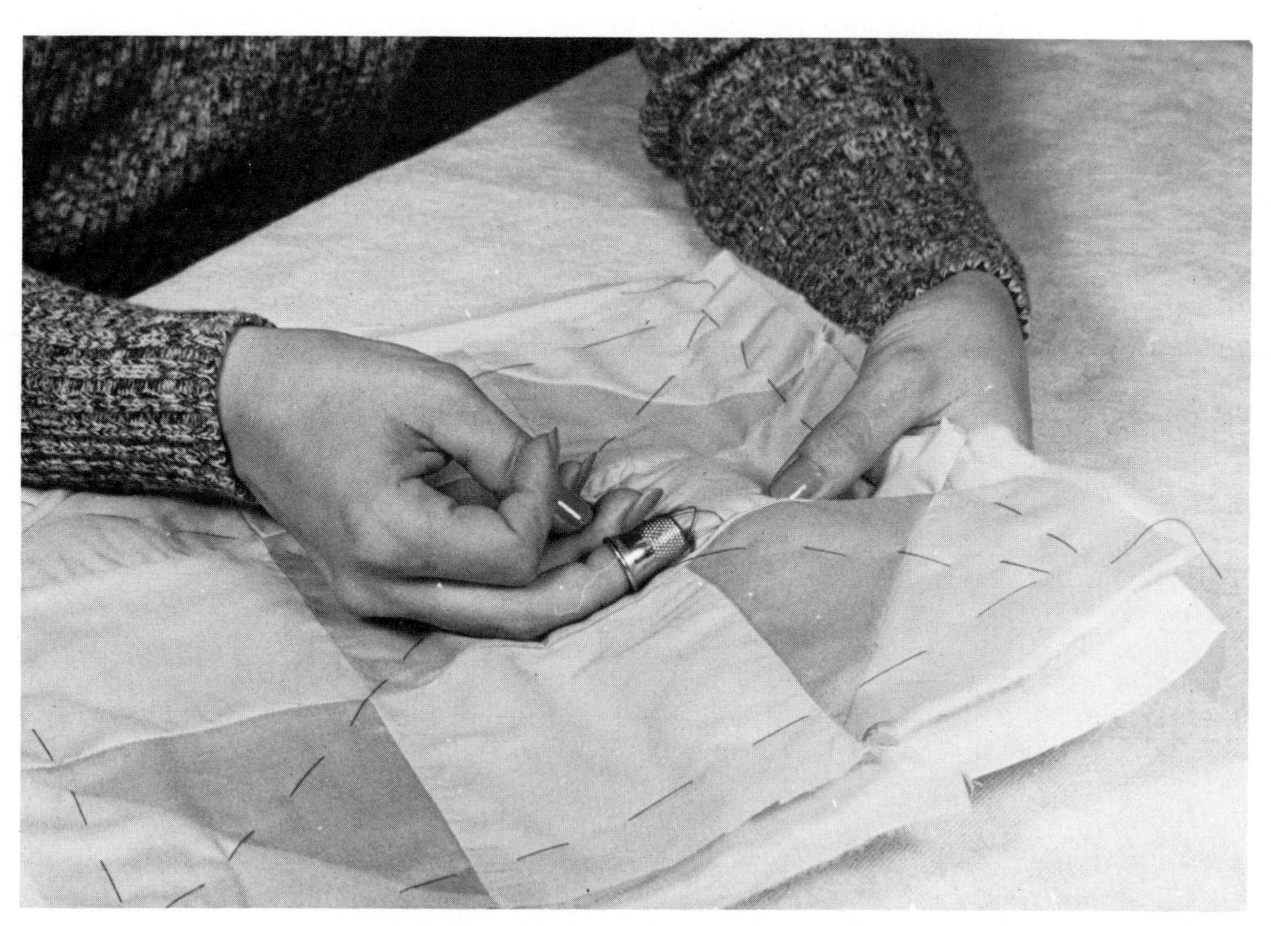

Quilting with a running stitch through all layers.

The needle I recommend is a No. 9 Sharps, a medium-length fine needle with a small eye. Quilting thread, a heavy-duty cotton thread that has been waxed, is preferred, but is not always available. Cotton thread is the next choice, with cotton-covered polyester core thread to be used if necessary. Quilting is always done with a single thread. Double thread weakens with use from the two threads rubbing against each other.

The thread should be cut about 18" long and a *single* knot tied in the end. All quilting begins and ends on top of the block. A few running stitches are taken and then increasing tension is put on the thread until the knot pops through the fabric and is caught in the batt. Sometimes it may be necessary to jerk the fabric away from the knot. This will not harm your quilt block.

Ending a row of quilting consists of tying a single knot, clipping the thread just above it, and gently popping it through the fabric. The easiest way to do this is to pull apart the top block from the backing fabric.

Stay at least one stitch away from any other quilting or the knot will not pop through. Some people find it easier to pop knots from the back. A knot should never show on the front or back of your quilting.

Quilting patterns vary from block to block. Some common and pleasing quilting designs for pieced blocks are illustrated below:

Parallel lines quilted on the bias.

Parallel lines quilted on the bias in both directions, forming the diamond pattern of quilting. *Block courtesy Dickie Schmidt.*

Each piece outlined ¼ inch inside the seam line.

Each light piece outlined ¼ inch inside seam line. This makes dark pieces puff up and become emphasized.

Appliquéd blocks are quilted differently. The first quilting done is to outline each appliqué piece in quilting stitches. This quilting should be done almost under each piece so that the stitches cannot be seen on the top. This makes each piece puff, making the block three-dimensional. This is often all the quilting that is necessary.

If more quilting is desired, the background square can be filled in as follows:

Parallel lines 1 inch to 3 inches apart, done on the bias.

Parallel lines 1 inch to 3 inches apart, done on the bias, in both directions.

"Luma-luma" quilting (Hawaiian wave quilting) consists of lines following the outline of the appliqué piece. These lines should be ¼ inch to 1 inch apart.

When quilting an appliqué block to be used in a quilt rather than a pillow, it is necessary to quilt a frame around the block, ½ inch to 3/4 inch inside the raw edge. This holds the batt in place when joining blocks.

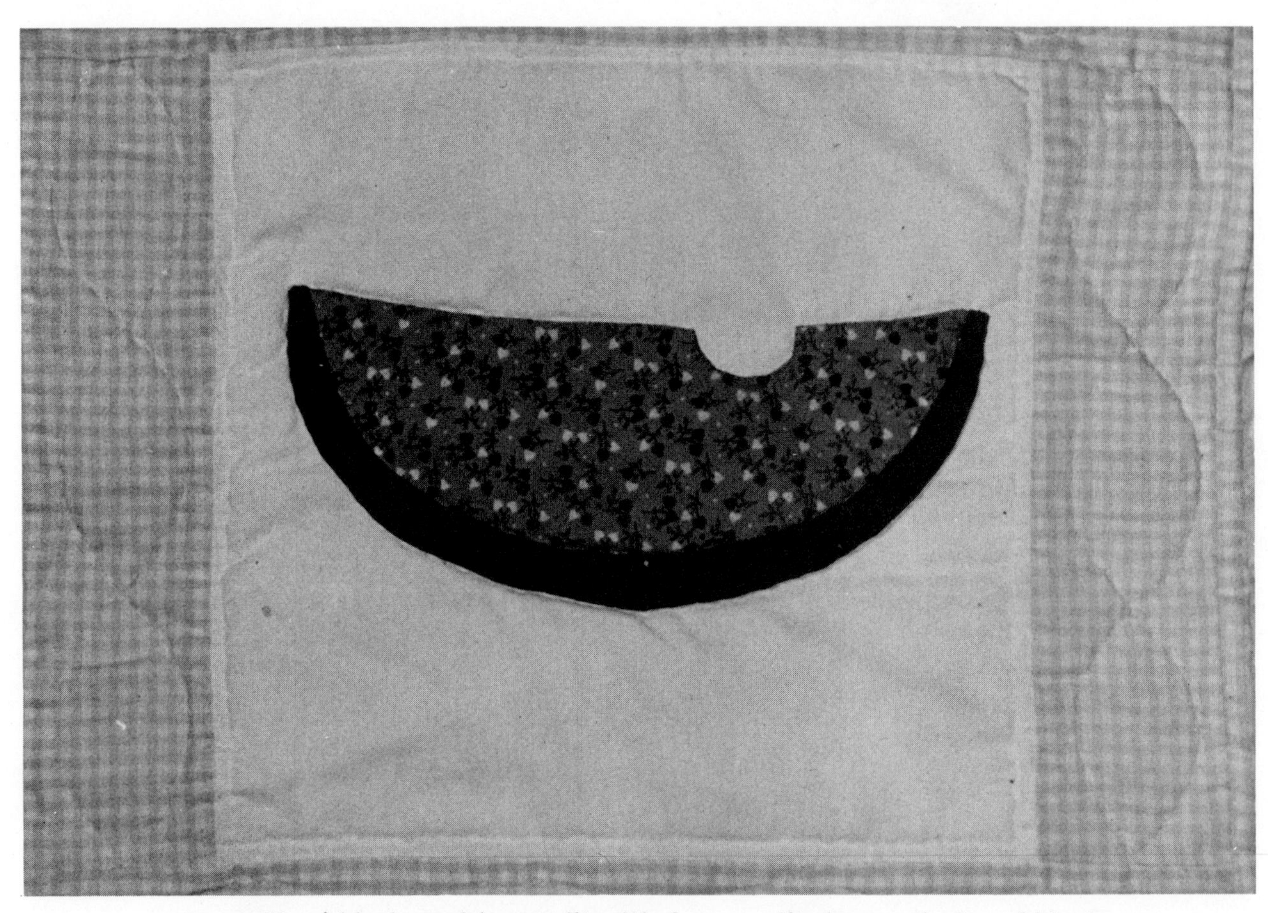

An appliqué block used in a quilt with frame quilted around edge of block.

Quilting stitches should be of uniform size on the top and bottom of the block, 12 to 16 to the inch. (Count stitches on the top and multiply by two.) The back should finish as smooth as the front.

Large areas (borders, etc.) look best when quilted on the bias. This seems to eliminate most gathers and puckers.

Quilting is sometimes done with a backstitch. This makes the play of light and shadow on the quilt top more pronounced, but is a great deal of work.

5

White-on-White Quilting

Although white-on-white or plain quilting can be used for an entire quilt, it is most often used in conjunction with pieced or appliqué patterns. Alternating a white-on-white block with a pieced or appliquéd block gives emphasis to the colored block by framing each one. White-on-white is also useful for borders and for decorative pillows for formal rooms.

White-on-white designs can be purchased from various sources and can be copied from books and magazine articles. Original designs are not difficult to draw using such items as saucers, tumblers, spoons, and cookie cutters.

A typical 10-inch quilt border can be quilted simply in white-on-white by drawing parallel lines on the straight grain of fabric about 2 inches apart. After quilting the borders and attaching them to the quilt, the lines of quilting are continued to form checkerboards, which add to the charm of the quilt.

Designs can be transferred to fabric in several ways. Straight lines can be drawn directly onto the cloth with a ruler. More complex designs can be traced onto the cloth, providing lightweight fabric is used. Place fabric square over the design and trace directly onto the fabric. Designs should be marked using well-sharpened soft pencil (very lightly), chalk, or a sliver of soap, depending upon the color of the fabric. Make sure the design to be traced is outlined in very black ink. If a commercial pattern is too light, go over it in black, felt-tip marker. Never use ballpoint pen to darken patterns, since it smears easily and can be transferred to the fabric.

If a dark, heavier-weight fabric or a fine fabric such as silk is used, tracing directly may not be possible. In these cases, I recommend tracing the design onto the backing fabric and quilting from the back. If that method is not satisfactory, the design should be transferred to tissue paper, which is basted to the top fabric. The three layers are then basted together and quilting proceeds as before. After quilting is finished, the tissue is torn away.

Generally, the thread color used for quilting white-on-white matches the fabric color. Sometimes contrasting thread can be very effective; however, care must be taken that quilting stitches are tiny and even, since they will be noticeable.

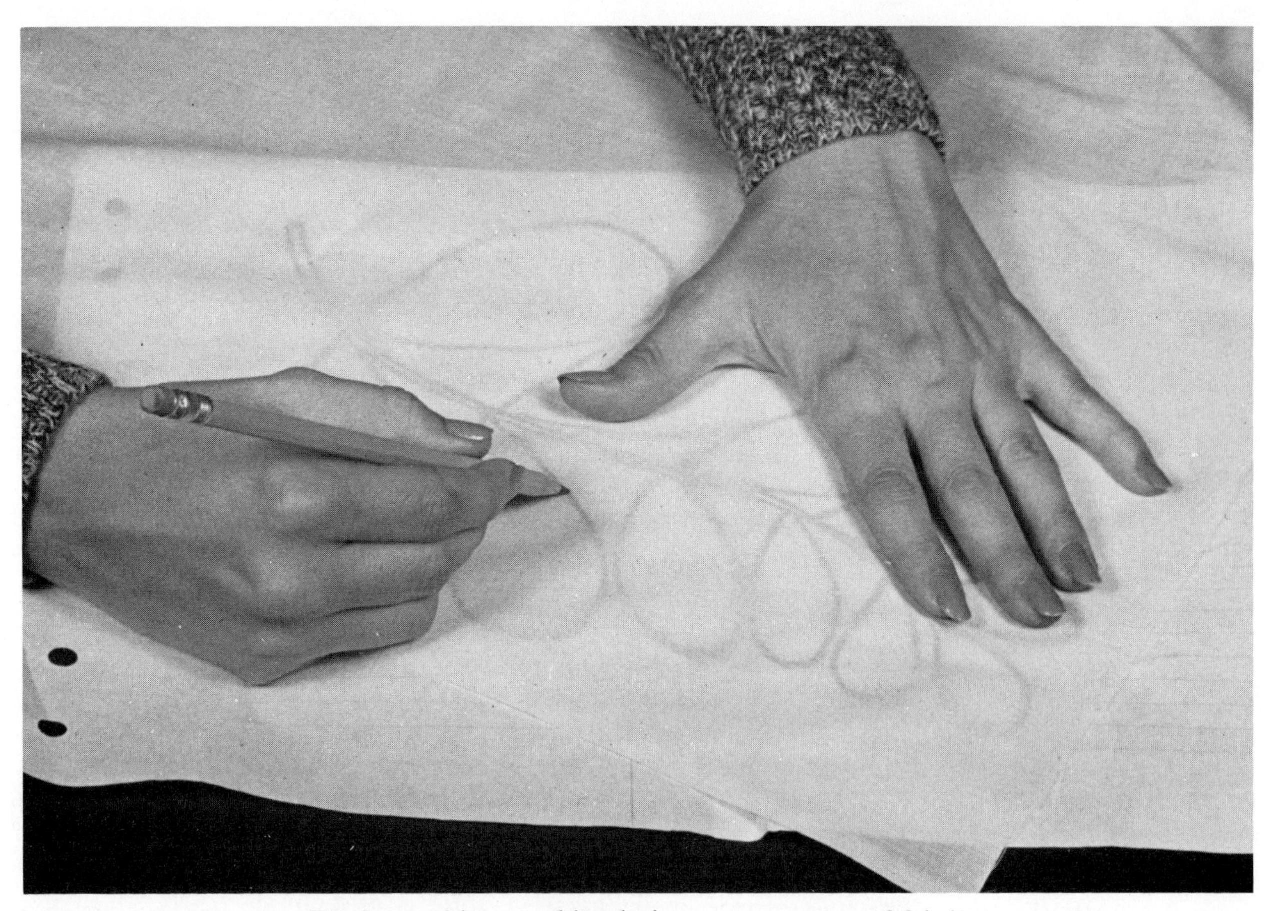

Tracing a white-on-white design onto a square of fabric.

6

Block Method Construction

I have always taught my classes block method quilting—working on one quilt block or strip at a time, never being overwhelmed by a huge piece of quilting to do.

When you are ready to join two quilted blocks, they should be positioned one on top of the other, *right* sides together. Pin, then machine stitch, the two top pieces together, using ¼-inch seams. *Do not* stitch through batt or backing fabric, only the fronts.

Next, lay the two blocks face down on a table top and carefully trim away the excess quilt batt, even with the seam lines.

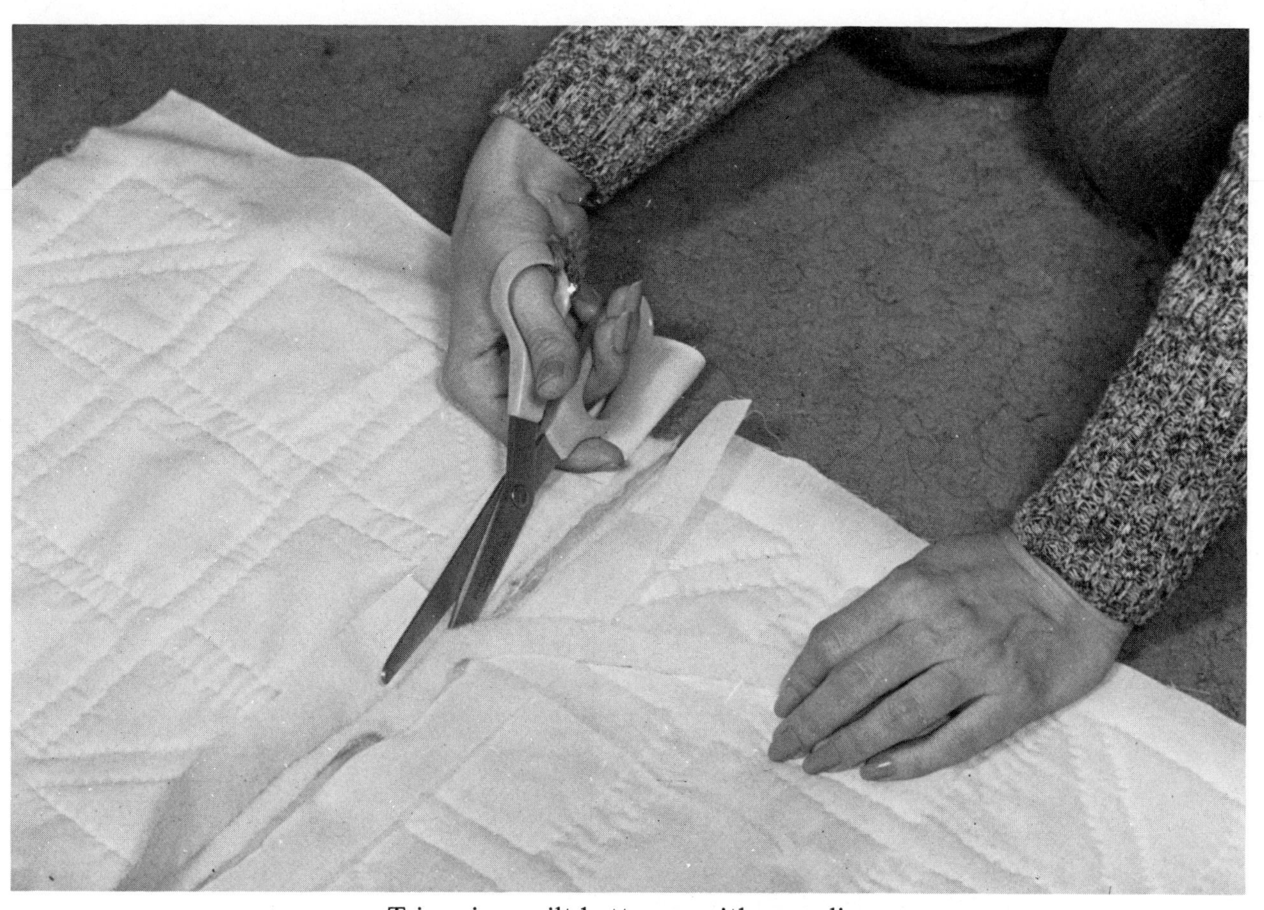

Trimming quilt batt even with seam line.

Then flatten one backing fabric piece and fold the other backing piece over it, turning under the raw edge approximately ¼ inch. These backing pieces should be trimmed if too large.

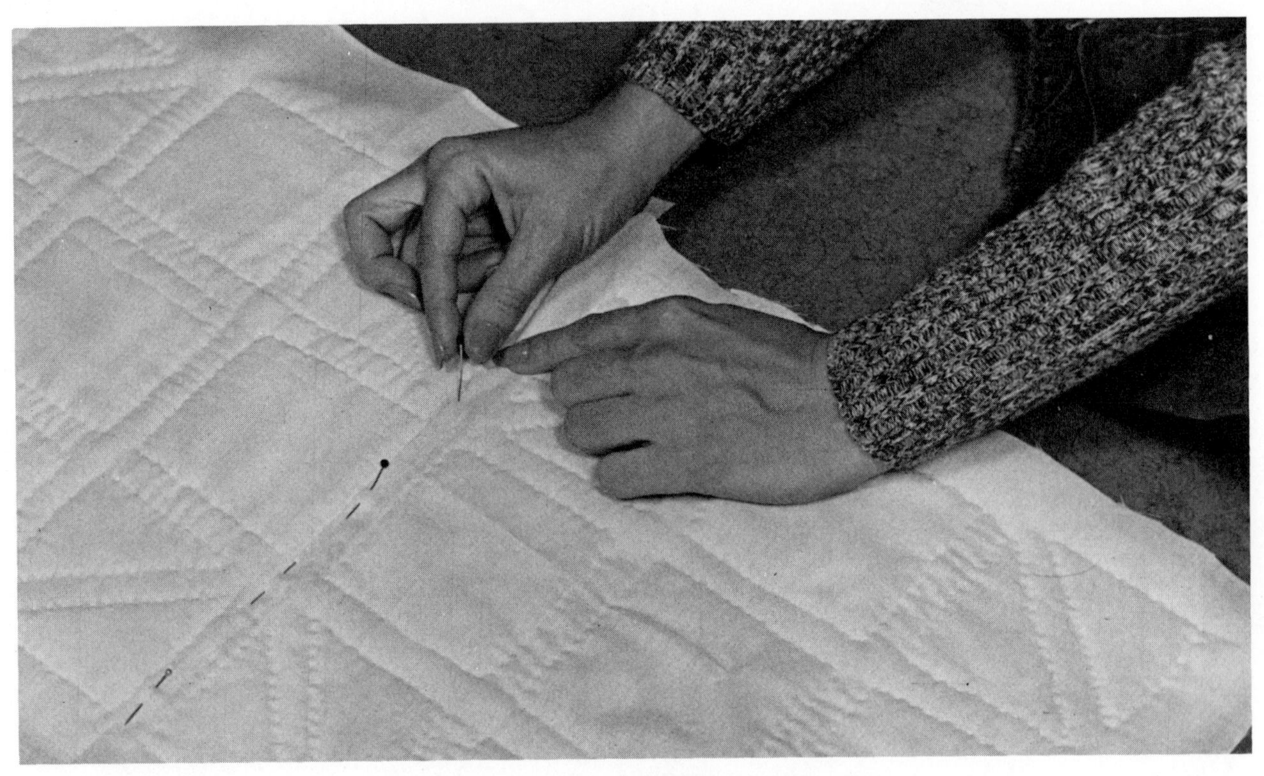

Folding and pinning backing fabric in place.

Pin in place and carefully blind stitch, making sure needle only goes through backing fabrics, not batt, and never shows on front of quilt. Fabric can be folded to the right or left. The blind stitched seams do not have to be consistently turned in one direction.

Quilts are assembled in rows, working widthwise across the bed. The completed rows are assembled by block method, just as if they were two blocks instead of two rows.

If a quilt is being made with lattice strips between blocks, the same principle applies to construction. Borders are added in the same way as lattice strips.

Borders are always quilted before assembly. Experience has taught me that borders should be cut at least one inch longer than the length of the quilt side they are to fit. For instance, if the side of the quilt is 95 inches long, cut the border the width desired and 96 inches long. Quilting seems to absorb the excess length.

Blind stitching the back of joined blocks.

It is easy to work block method in rows of blocks when making a quilt. This actually saves time because it reduces hand work. If a quilt is to consist of 42 blocks, 6 across the bed by 7 down the length, piece, then sew together, 7 rows of 6 blocks each. Then put your layers together with your batt and backing fabric in a long strip instead of a smaller square. Pin layers together and begin quilting at one end, working toward the other end and adjusting pins as you go to keep the back taut. Eventually your quilt will be assembled with only 6 machine-sewn seams and 6 blind-stitched seams.

7 Binding

Binding and finishing a quilt begins with the preparation of narrow strips of fabric, seamed together to equal the circumference of the quilt, plus 12 to 24 inches to allow for mistakes. These strips should be approximately 1½ to 3 inches in width and can be cut or torn on the straight grain of fabric. Bias strips for binding are needed only for binding scalloped-edge quilts.

Binding is placed against the edge of quilt front, right sides together, and sewn by machine using a ¼- to ½-inch seam.

Corners can be handled in several ways, the easiest approach being a "false mitre" shown below:

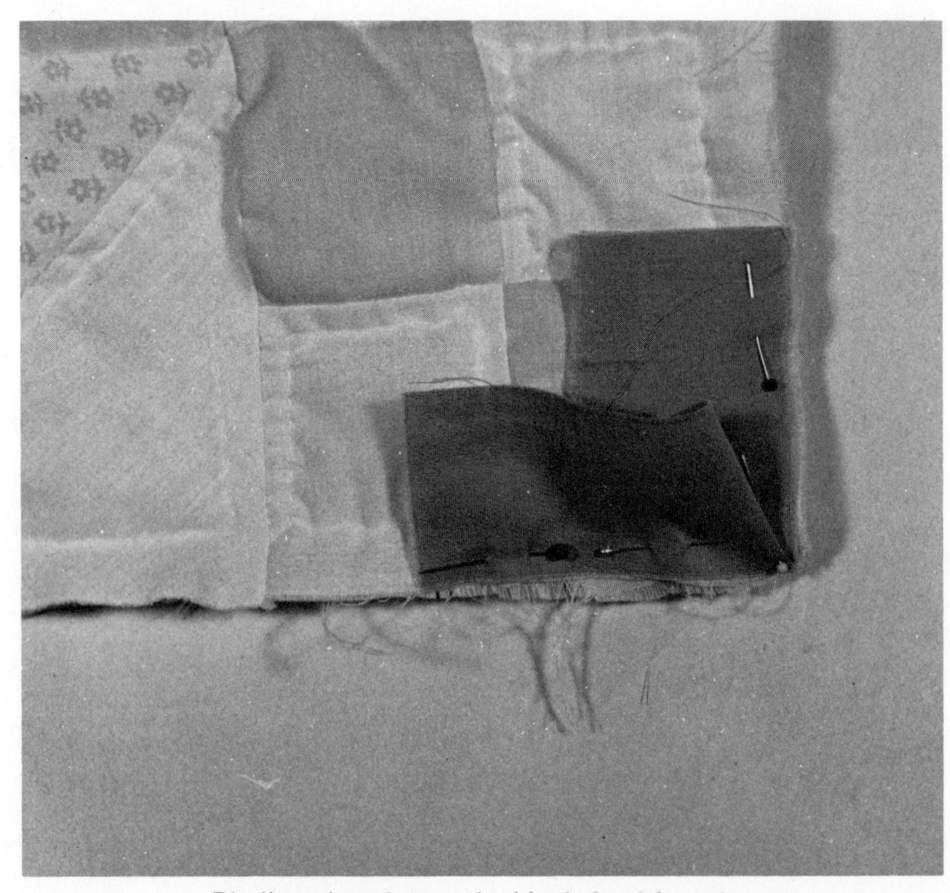

Binding pinned onto the block for false mitre.

Corners of the quilt can also be gently rounded and the binding eased around the curve.

Binding pinned for rounded corner.

Corners can also be treated by starting and ending at each. Sew binding along length of one edge. Turn binding out from edge of quilt and apply binding to next edge, leaving about 1 inch of excess binding at beginning. See below for further instructions.

When all machine stitching is complete, turn binding over raw edge, and blind stitch to quilt backing.

The beginning and ending of binding should be done as follows:

Front view of binding showing overlap required.

Back view of corner, showing first piece of binding pinned in place.

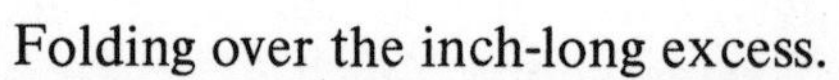

Folding over the inch-long excess.

Raw edges of second piece turned under and pinned down.

Occasionally, through faulty arithmetic, a finished quilt will be too narrow. In such a case, a stuffed binding will add up to 2 inches to each side.

Stuffed bindings should be cut twice the desired finished width plus ½ inch. For example, for a finished stuffed binding of 2 inches, cut strips 4½ inches wide. The binding is applied the same way as a regular binding, except that before hemming down the back, stuffing of quilt batt strips is inserted.

A quilt should always be signed and dated. Initials and the year are enough. They can be embroidered on a block before quilting or quilted in a block or along a border.

8

Tabletop Quilting

If a pieced or appliquéd quilt top has been acquired and it cannot be cut into sections for quilting by the block method, it can still be quilted without a frame.

A backing slightly larger than the finished top should be prepared. Polyester and cotton blend fabric can be seamed to make a piece large enough. Bed sheets should *not* be used. They are too hard to get a needle through.

The finished backing is laid on a floor (seams up), then a sheet of quilt batt carefully placed on top of it. Check frequently to be sure quilt back remains smooth. The completed quilt top is positioned on top of batt, right side up, and the quilt basted as follows:

Begin in the center and baste to each corner, then baste parallel lines down the length of the quilt 8 to 12 inches apart.

The quilt should be placed on a table for ease in quilting. Stitching should begin in the center and continue to the edges. In this manner any unevenness is worked out and the back remains smooth.

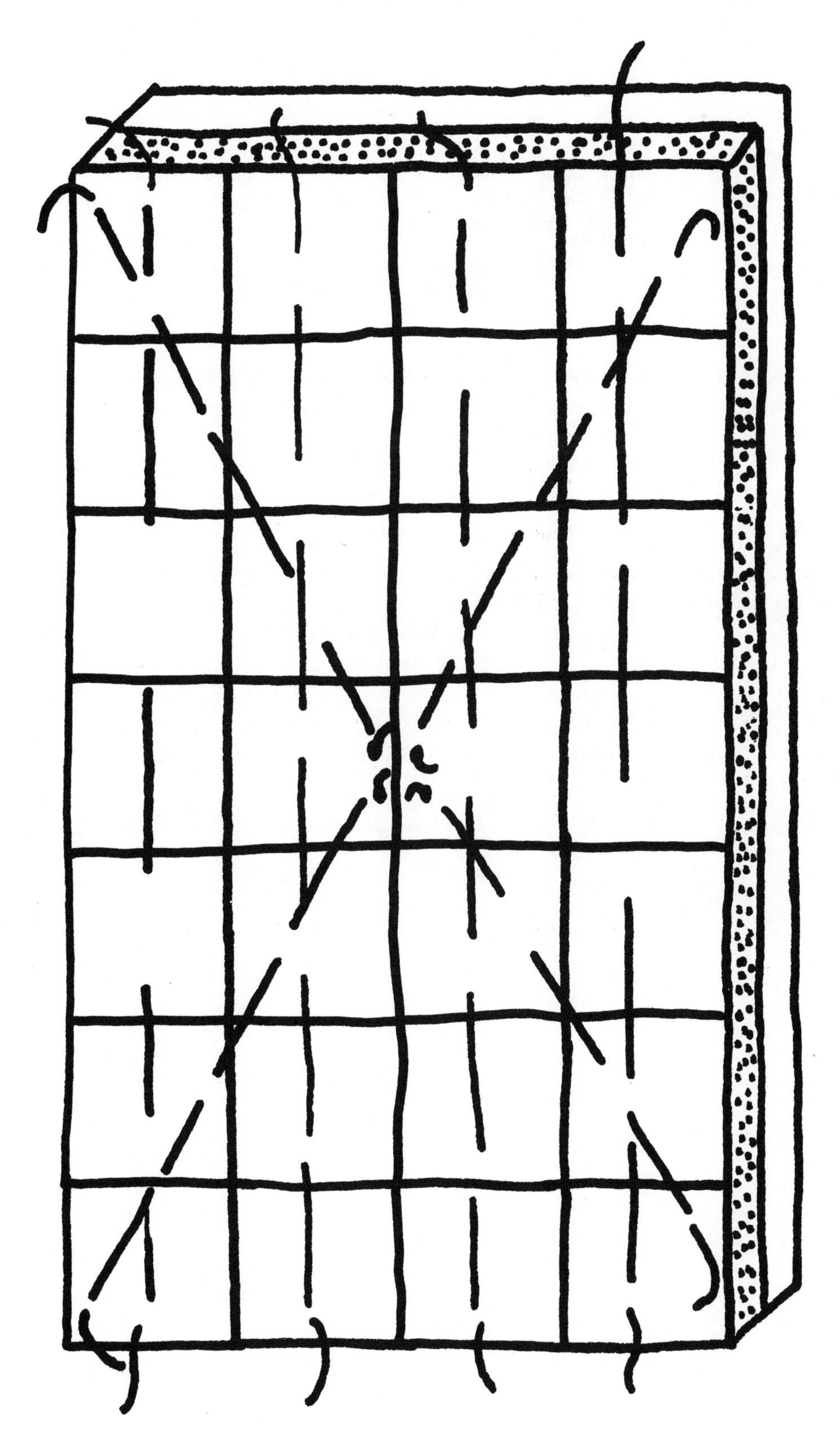

Quilt basted ready for tabletop quilting.

9 Quilt Planning

Typically, quilts are made coverlet size, and can be used as bedspreads with a dust ruffle. Minimum dimensions in inches are as follows:

Crib	36 x 50
Single	60 x 95
Double	75 x 95
Queen	80 x 100
King	100 x 100

Some latitude is allowed in length and width, since a quilt can be overlarge for the bed it is to fit and still look well.

If a quilt is to be made to fit to the floor, careful measurements of the drop, the length from the edge of the mattress to the floor, must be made. For example, if the bed involved is a queen size with a drop of 22 inches, the width of the quilt should be 104 inches (60 inches plus 22 inches plus 22 inches). The length should be 112 inches (80 inches plus 22 inches plus 10 inches.)

Before fabric is purchased, the quilt should be sketched with colors indicated as well as dimensions and design.

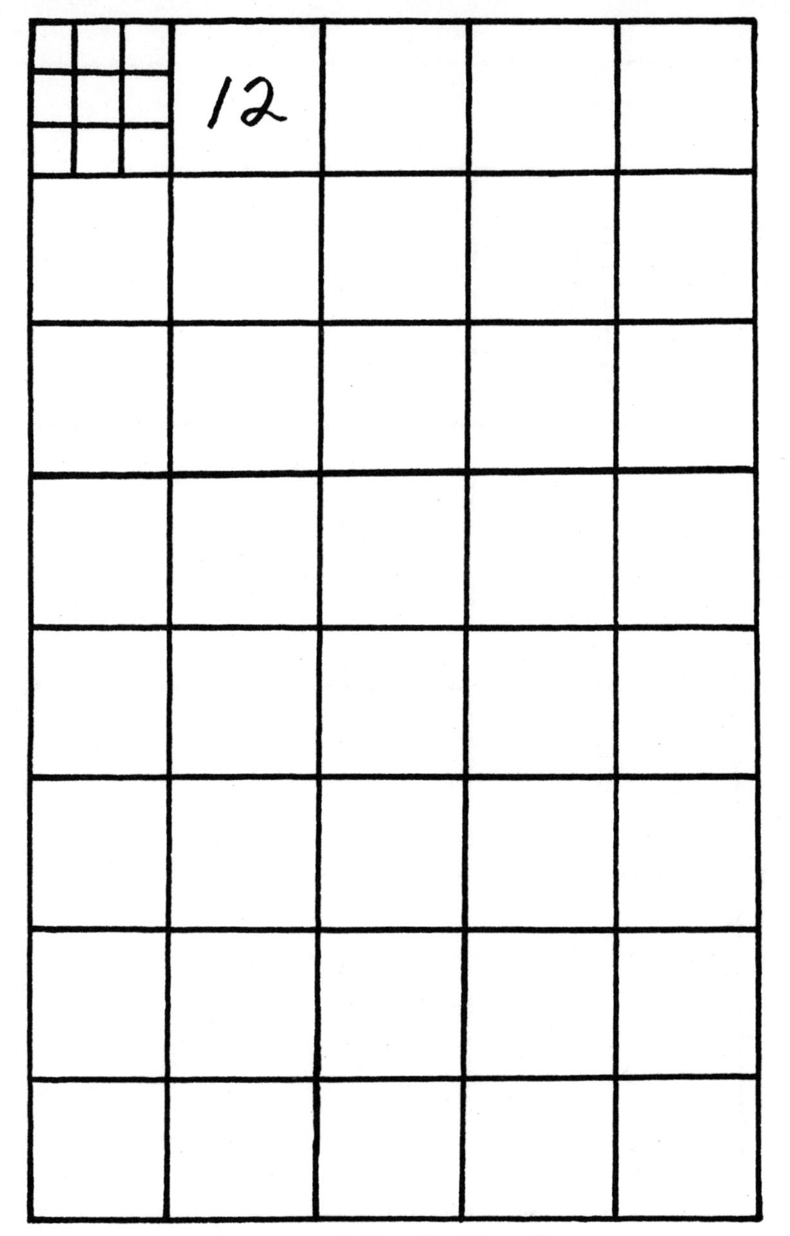

Sketch of Nine-Patch quilt.

It is useful to keep records of all data involved in each quilt made, since another quilt like it could be made at some future time. Included should be amounts of materials purchased and price, overall finished dimensions, notes of mistakes made and corrected, etc.

The amount of fabric to buy is difficult to determine. An average pieced quilt top, double-bed size, requires about 5 yards of fabric for the top alone. If the design consists of equal amounts of light and dark fabric, purchase 2½ yards dark and 2½ yards light. To be safe, cut out all pieces needed for quilt at once and then purchase more immediately, if needed.

A sketch for a more complicated design follows:

Sketch of Shoo-Fly quilt.

10

Hawaiian Quilting

Hawaiian quilts are different from our other American quilts. When the missionaries brought quilting to Hawaii in the nineteenth century, the native Hawaiians had no old clothes to cut up for pieced or appliqué work. All new fabric was used and developed by the natives into Hawaiian quilting as we know it today.

Hawaiian quilts have many traditions and superstitions attached to them. For instance, it was considered bad luck to copy someone else's pattern, and also unlucky to quilt them while seated on the ground. In the beginning, many were made in the royal colors of red and yellow, but no matter what colors were used, they were always solids.

Traditional Hawaiian quilts are made of one central appliqué motif done in the cut-paper method. That is, fabric for the appliqué is folded several times and the design cut just as we cut paper snowflakes as children. The borders and corners of Hawaiian quilts can be filled in with portions of the central appliqué design or other cut-paper designs, or the remainder of the quilt can be left plain.

Very attractive pillows can be made from Hawaiian designs. Even the simplest design is effective.

All Hawaiian appliqué should be pinned securely to the background fabric and appliquéd with the blind stitch. Since Hawaiian patterns have extremely narrow indentations, they are appliquéd differently. When at least 1/8 inch of fabric can no longer be turned under, begin appliquéing with a satin stitch, narrow at first, then widening as the point of the indentation is reached. The satin stitch should narrow again after this point is reached. A crescent-shaped piece of satin stitch is ideal. For the finest effect, thread color should match the appliqué piece. As always, avoid polyester thread for hand work.

The inside of the appliqué piece as well as the background fabric is always quilted in luma-luma, Hawaiian wave quilting. This quilting follows the outline of the appliqué piece. When appliqué is complete and you are ready to begin quilting, it is helpful to mark the appliqué piece for quilting, lightly, with chalk, pencil, or a soap sliver. Then begin your quilting in the center of the appliqué and work out. There is no need to mark lines for quilting on the background fabric. Quilting "by eye" is accurate enough. The lines of quilting should be ¼ inch to 1 inch apart.

Hawaiian pattern appliquéd with blind stitch and satin stitch. *Block courtesy Charlotte Healey.*

11 Novelties

Among traditionalist quilters, novelties are not much done. But some are so easy, so useful, or so much fun that I feel you should try each of them.

The simplest novelty is OUTLINE QUILTING. A printed fabric is used for the top of the article and is layered with batt and backing, and is basted in the usual way. Then each part of the design to be emphasized is quilted around. No pencil lines are necessary.

STUFFED APPLIQUÉ is also very simple. Each piece of appliqué to be emphasized is appliquéd, leaving a small opening unstitched. The piece is then lightly stuffed with loose polyester filling or quilt batt trims and appliqué stitching is completed.

Stuffed appliqué trim on a jacket back.

The YO-YO is a novelty practiced for at least the last one-hundred years. First step is cutting a circle of fabric, then folding over the edge ¼ inch and sewing it down with a running stitch. The beginning stitch should be backstitched or have a very secure knot. When running stitches are completed, gather the circle, forming a yo-yo. Knot remaining end of thread securely. A single or double thread may be used. These yo-yos are tacked together at the edges to form entire coverlets or pillow covers. The side of the yo-yo with the hole is the right side. I have seen other articles made of yo-yos, most memorable being an overskirt for evening wear and a window treatment in an extremely sophisticated apartment.

TRAPUNTO, or Italian Stuffed Quilting, is in its traditional form done as follows: two fabrics are basted together (no batt is used) and designs quilted on it such as fruit, leaves, or vines. Then the threads of the backing fabric are gently separated and stuffing is inserted to puff each part of the design. When stuffing is completed, the fabric is smoothed together on the back. As you can imagine, this requires tremendous amounts of time.

Trapunto for such items as pillows and clothing can be done in a much easier manner. The old method is followed until the point of stuffing is reached. Then you should *carefully* make small slits in the backing fabric and insert bits of stuffing. When stuffing is complete, the slit is whipped closed. Trapunto is beautiful when done on printed fabrics also. Parts to be emphasized are quilted around, then stuffed.

Back view of trapunto showing slits that have been whipped together.

Trapunto quilts, or other items where the back may be seen, should be lined.

The BISCUIT is one of the nicest novelties and perhaps the most fun of all. Every stitch is done by machine so the finished article is extremely sturdy. Two pattern pieces in cardboard should be prepared for each biscuit size. Typical biscuits are:

To Finish	Cut Bottom	Cut Top
1 1/2"	2"	3"
3"	3 1/2"	4 1/2"
4"	4 1/2"	6 1/2"
6"	6 1/2"	9 1/2"

The top square should be of the fabric you want to show. The bottom square is never seen and can be made of old sheets, extra fabric remnants, etc.

The biscuit is assembled as follows:

The top square is placed right side of fabric up on top of bottom square. Match corners of both squares on one side and put a pleat in top fabric to take up extra fabric, in about the middle, and pin.

Continue with sides two and three.

Stuff through open side, match corners, and pin. Amount of stuffing (loose polyester fiber) depends upon personal taste. Experiment.

Machine stitch around all four sides using between 1/8- and ¼-inch seam (distance from raw edge).

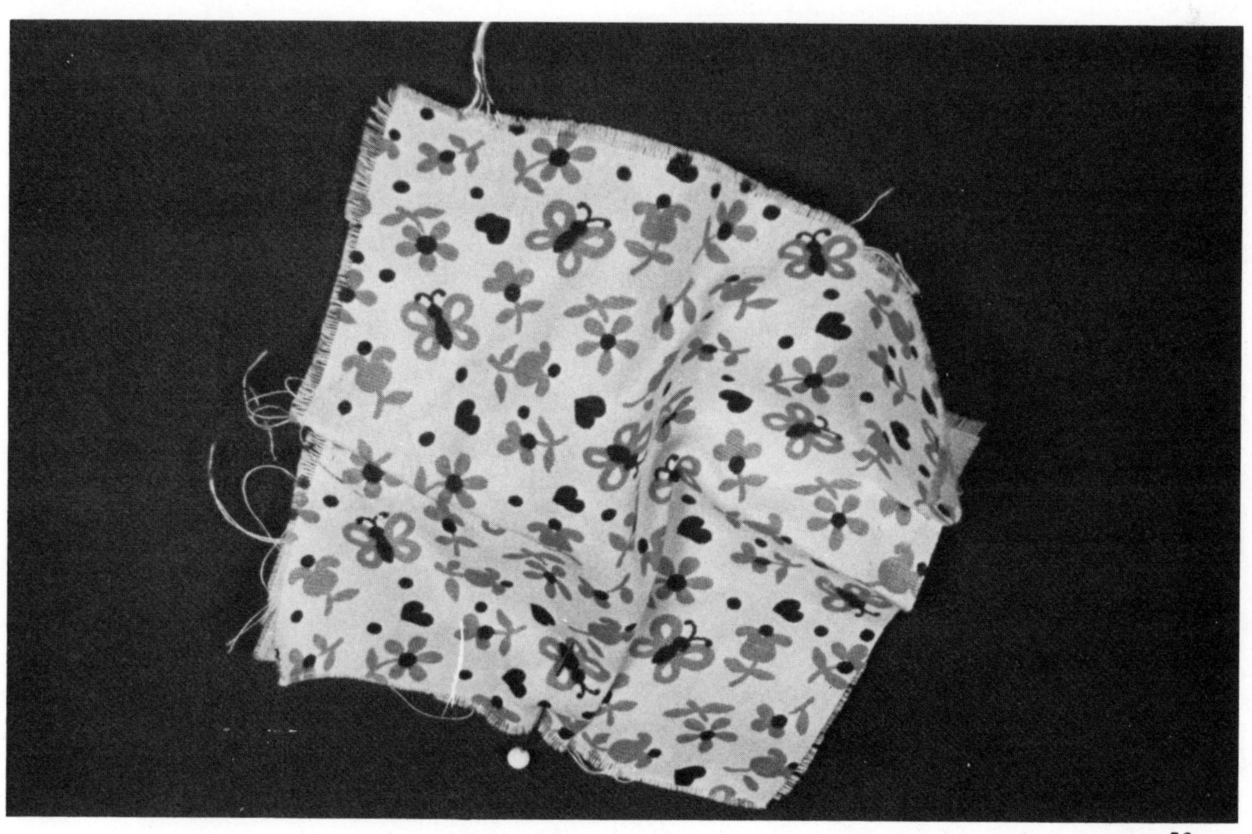

Biscuit with side No. 1 pinned.

Sides 2 and 3 have been pinned. Stuffing is being inserted through side 4.

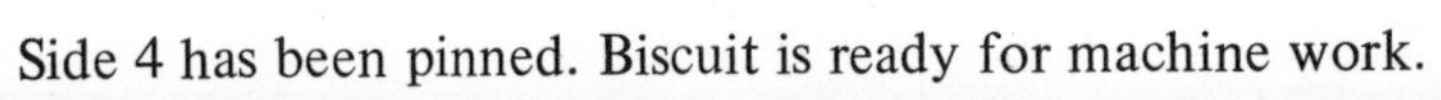

Side 4 has been pinned. Biscuit is ready for machine work.

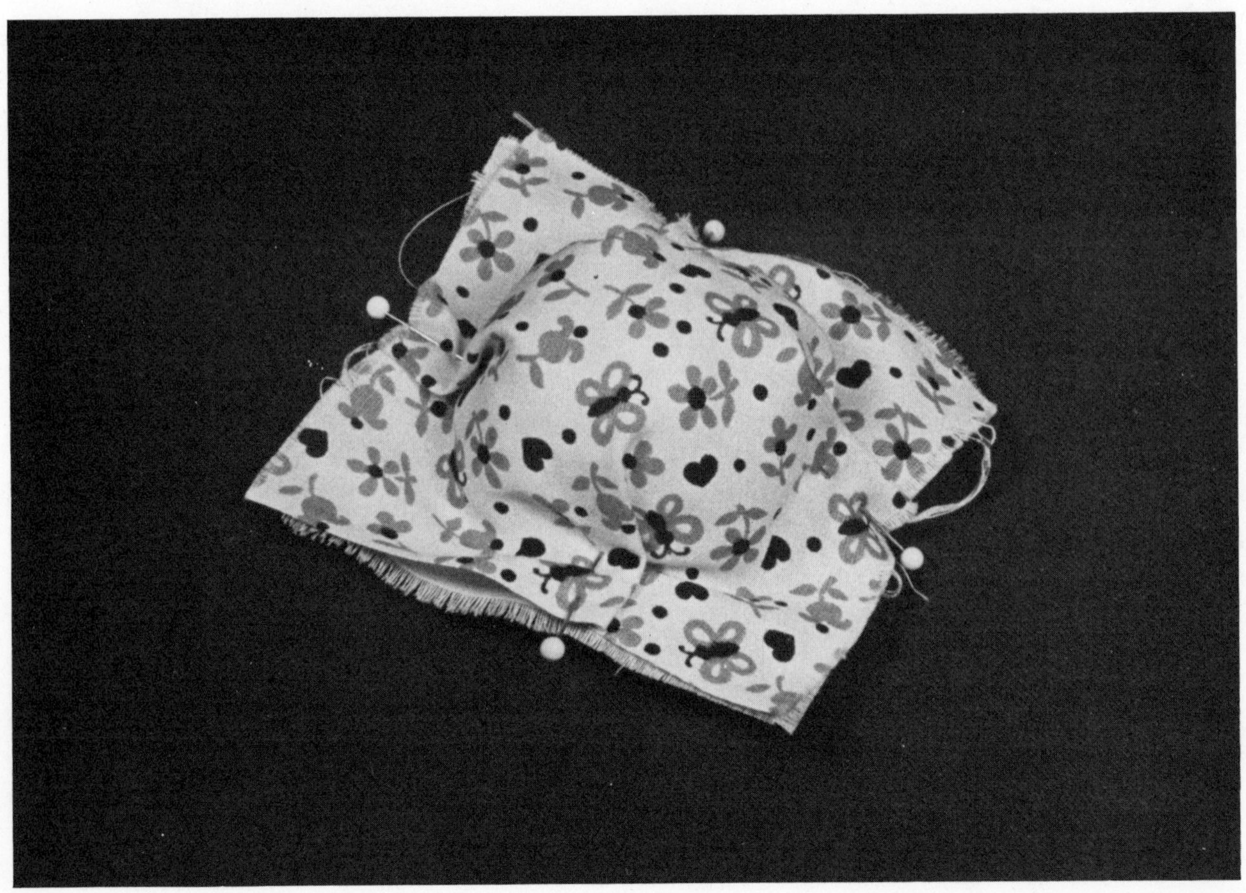

Twenty-five biscuits have been seamed together in 5 rows of 5 each.

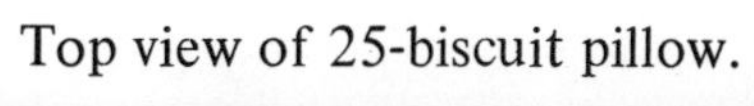

Top view of 25-biscuit pillow.

Sew completed biscuits into rows, using a bit more than ¼-inch seam. Check frequently to make sure first stitching does not show. Completed rows should be sewn together with same seam as above. As always when piecing rows into a block, match seams carefully.

When making biscuits consisting of squares of 1-inch difference, one pleat in each side is sufficient. When there is a 2-inch difference, 2 pleats, evenly spaced, are used. A 3-inch difference requires 3 pleats per side, etc.

Biscuits are usually lined and made into coverlets and crib quilts, but pillows, chair pads, and bench pads are all effective.

12 Restorations

Antique quilts, whether inherited or purchased, often are in poor condition. These can be restored with patience and appliqué.

Whether the quilt is a pieced design or an appliqué, worn pieces are replaced by appliquéing new fabric directly over the worn ones.

Fabric chosen for restoration should be the same *type* of fabric used in the original (i.e., cotton, silk, wool, etc.). If a tiny red and white print cotton fabric was used in the original, find a tiny red and white print cotton for the restoration. The prints will not be identical, but that will not matter.

Old fabrics are usually very faded. The replacement patch should be faded also. Sometimes it is possible to use the wrong side of modern fabrics to achieve this faded look. Usually fabrics must have color removed, using bleach or color remover.

After fabric is prepared, pieces ¼ inch larger than worn patches are cut, then appliquéd over worn pieces. If original batt has come out, replace with similar batt (cotton, wool, flannel). Quilting stitches should be redone on the replaced pieces.

Binding can be replaced with new binding in the usual way. It is not necessary to remove old binding.

Small moth holes can be carefully darned using silk floss or wool crewel yarn.

Restoration in progress showing 2½ triangles replaced in pattern.

Completed Designs

Antique pieced quilt in the Rainbow pattern. Each stitch was done by hand. Unprocessed cotton was used as the filler.

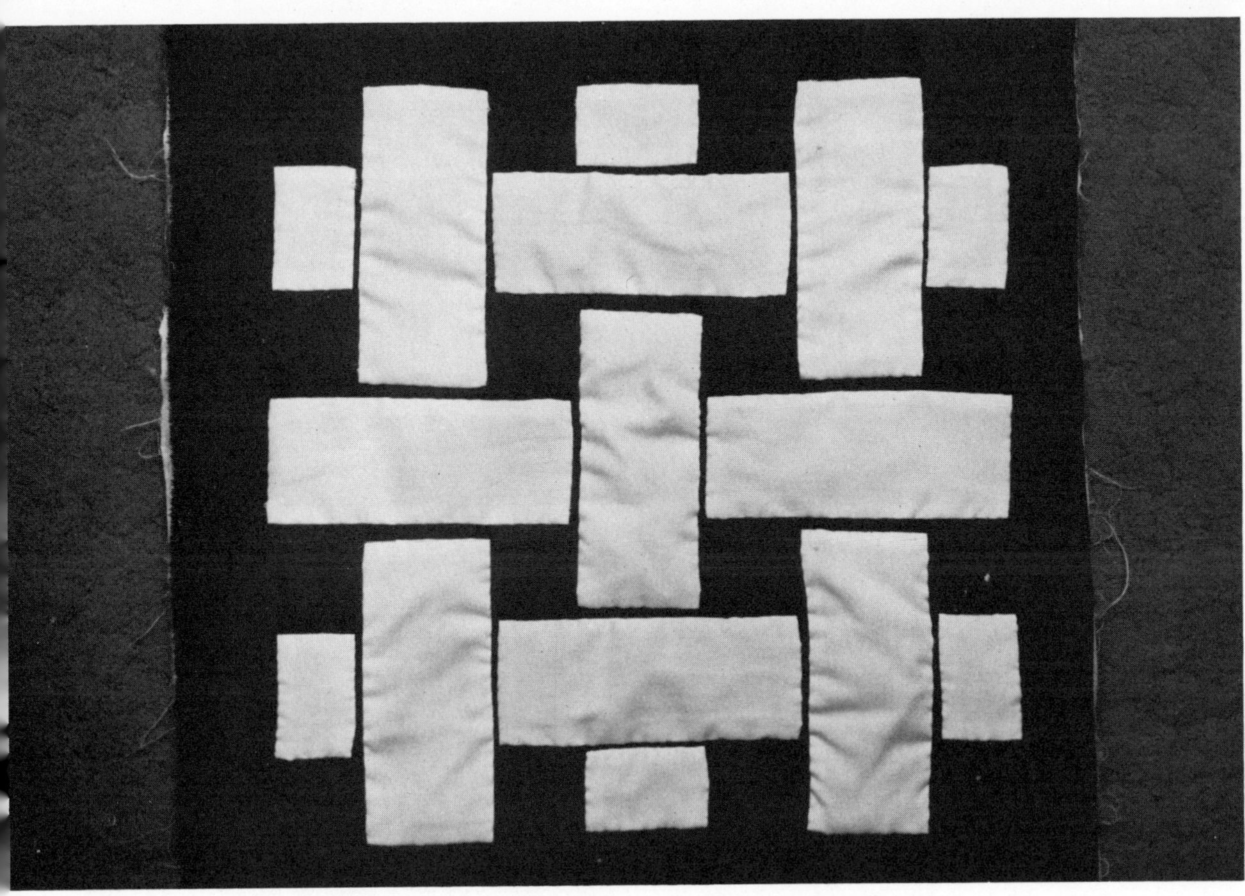

Basketweave appliqué block. *Made by Elaine Rosen.*

Crosspatch quilt. This was the second quilt I ever made. It has a very unsatisfactory cotton batt.

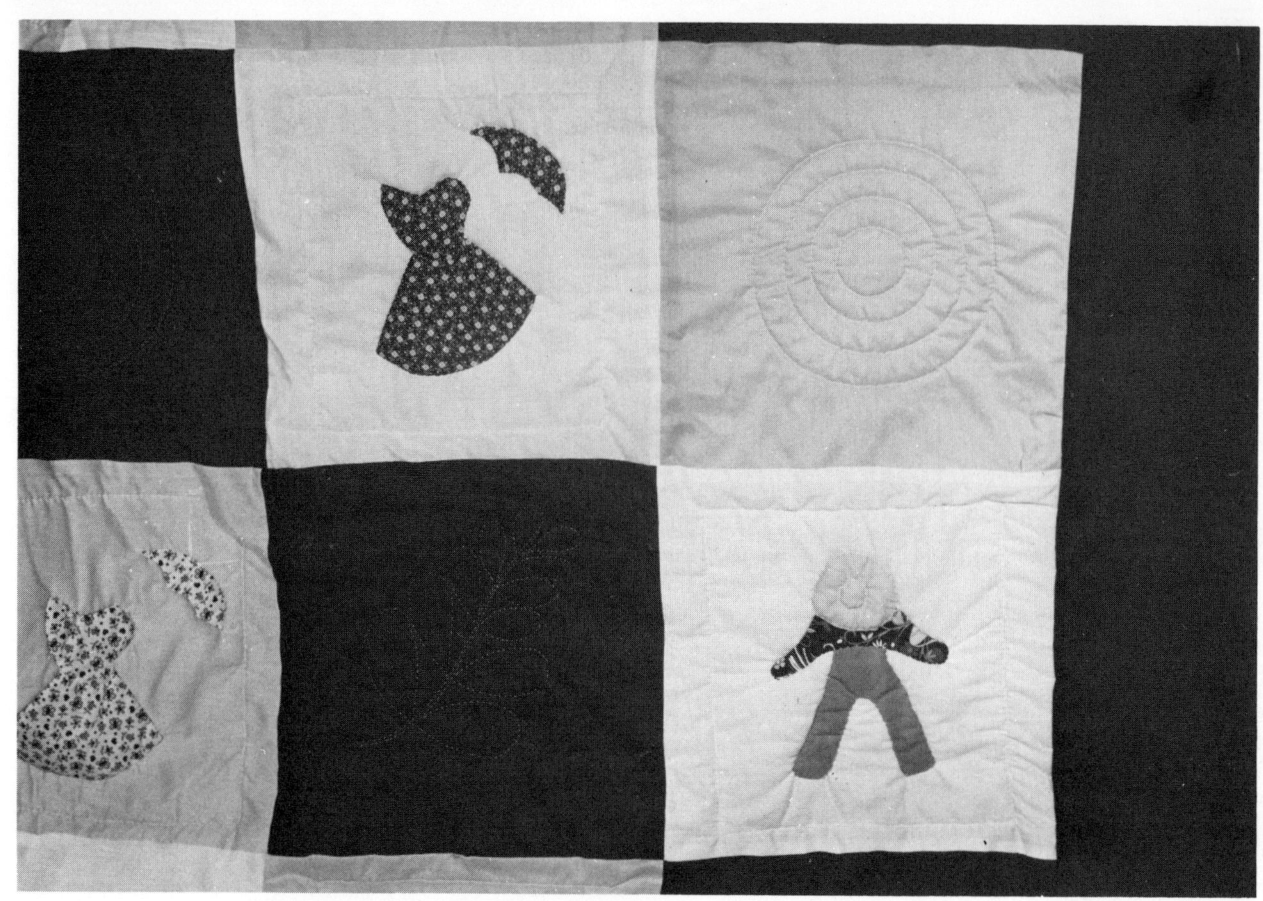

Umbrella Girl and Overall Bill quilt. *Made by Vicky Campion.*

Shoo-Fly crib quilt and pillow. *Made by Ann Eutermarks.*

Pieced and appliquéd quilt top, unquilted, in the Poplar Leaf pattern. This pattern is named for the tulip poplar leaf, which it resembles.

Snowflake crib quilt. *Made by Vicky Campion.*

Star Puzzle and Biscuit pillows. *Made by Bev Hensley and Marcia Forrest.*

Weathervane block.

Lincoln's Platform quilt. *Made by Marcia Forrest.*

Star Puzzle crib quilt.

Patterns

Many of my favorite traditional pieced blocks have been included in this section. I have chosen simple designs that work up into interesting quilts. All can be made as scrap quilts using many different prints and solids, or can be made with only two or three fabrics. Before attempting a quilt of any of these patterns, I recommend first making a sample block, then a sketch of the completed quilt. Experiment in your sketch with block-to-block-to-block layouts, alternating pieced and white-on-white layouts, and the use of lattice stripping before deciding on the final design. All of the pieced patterns have ¼-inch seam allowances already included.

The appliqué patterns can be combined many different ways to form blocks. All are drawn actual finished size. A 15-inch square is an excellent background block.

The white-on-white designs can also be combined for different effects. Again, experiment on paper before beginning any large project.

Album quilt.

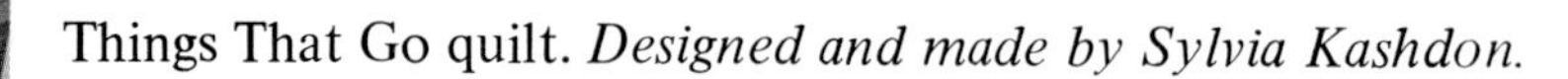

Things That Go quilt. *Designed and made by Sylvia Kashdon.*

Detail of Things That Go.

Detail of Things That Go.

Frog. *Designed and made by Marie Molokie.*

Apple and Mushroom pillows. *Designed and made by Geri Bring.*

Modern Geometric pillow. *Made by Elaine Rosen.*

Snowflake quilt. *Made by Diana Deal.*

Noah's Ark Wallhanging. *Designed and made by Pat Peters.*

Nine-Patch block. Cut 4 light and 5 dark. Block finishes approximately 15 inches square.

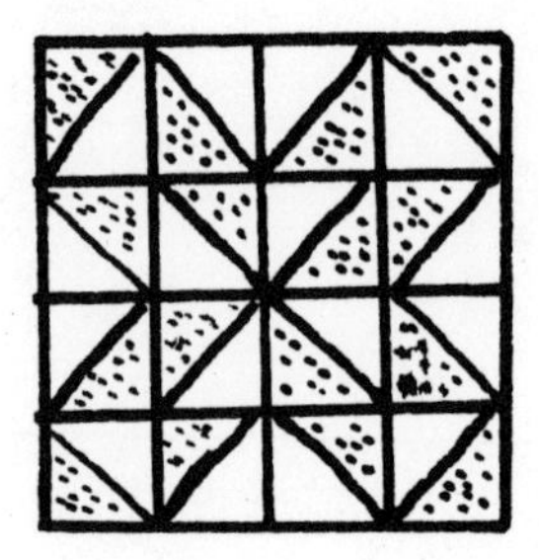

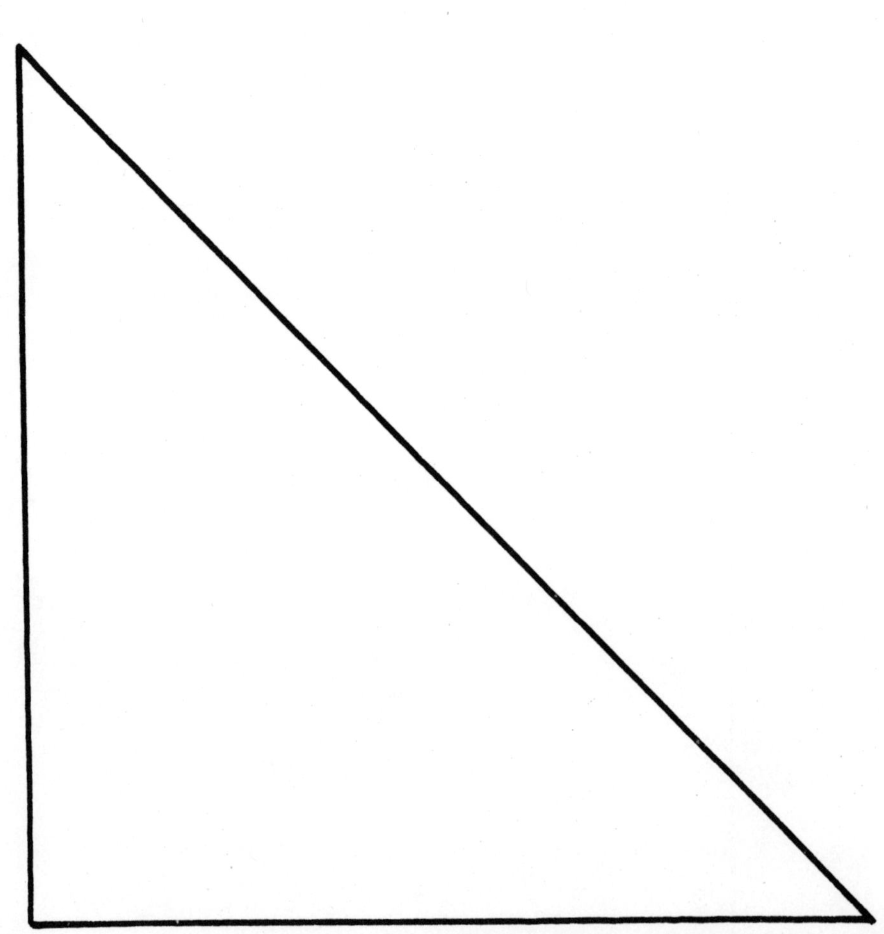

Star Puzzle block. Cut 16 light and 16 dark. Block finishes approximately 16 inches square.

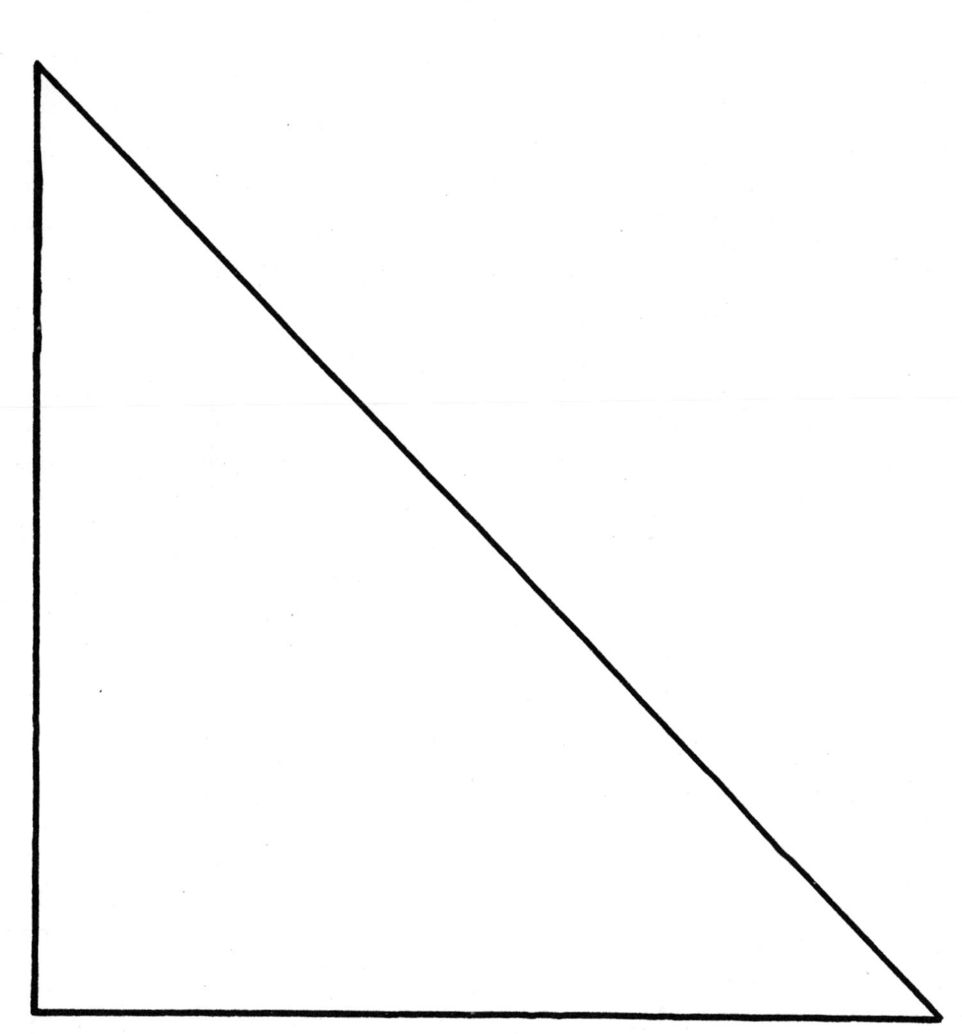

Hourglass block. Cut 8 dark and 8 light. Block finishes approximately 8 inches square.

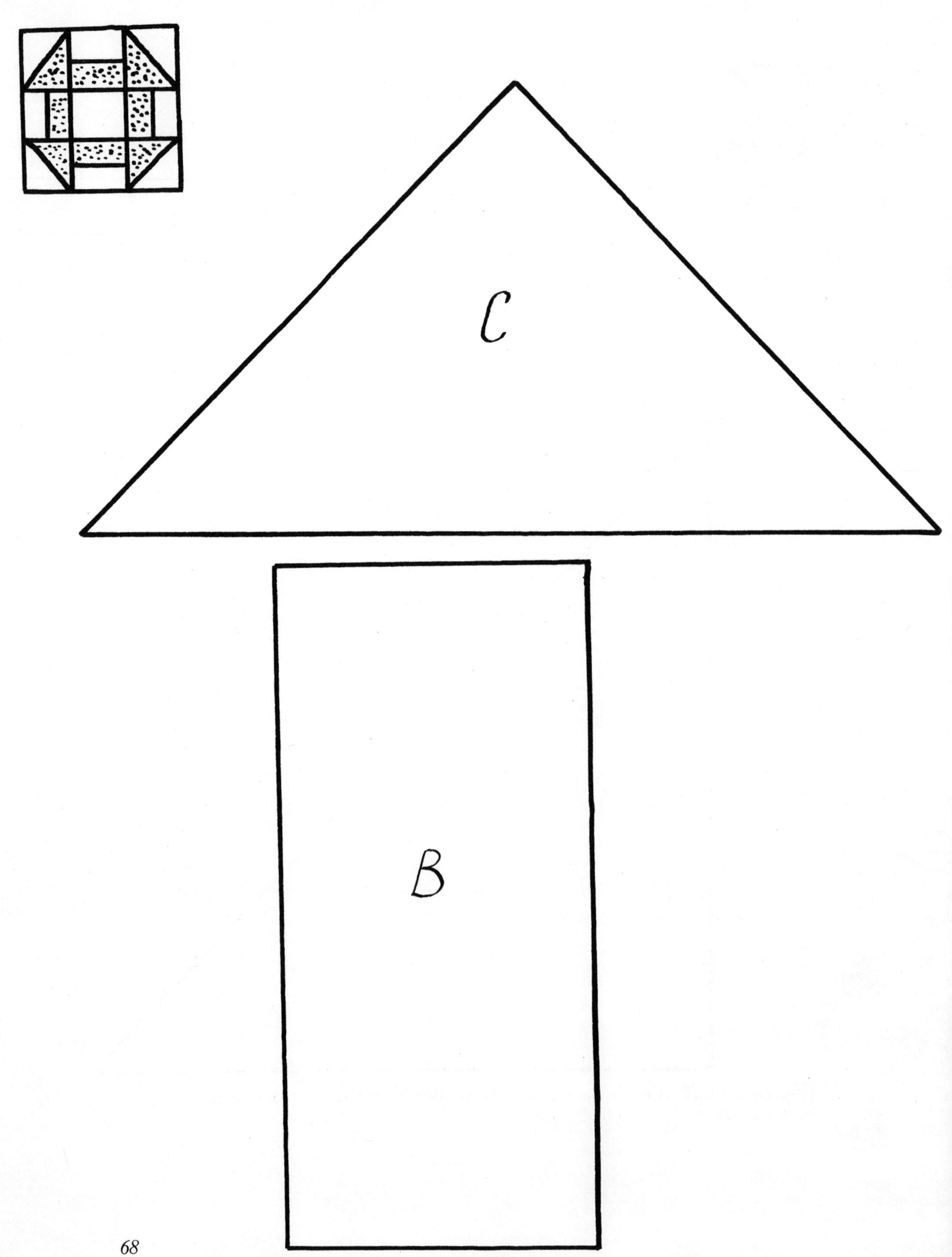

Churn Dash block. Cut 1 light A, 4 light and 4 dark B, and 4 light and 4 dark C. Block finishes approximately 12 inches square.

A

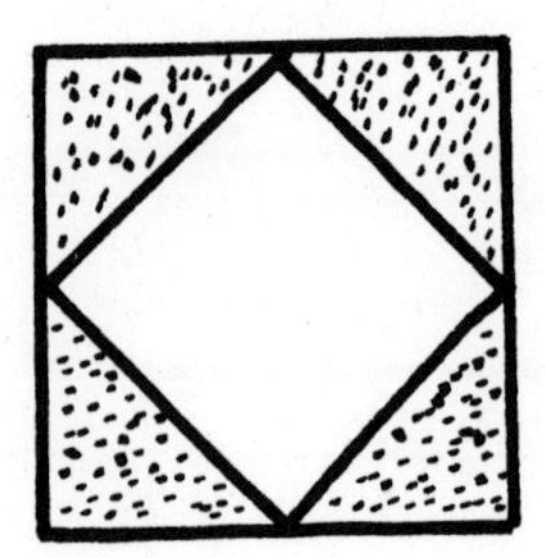

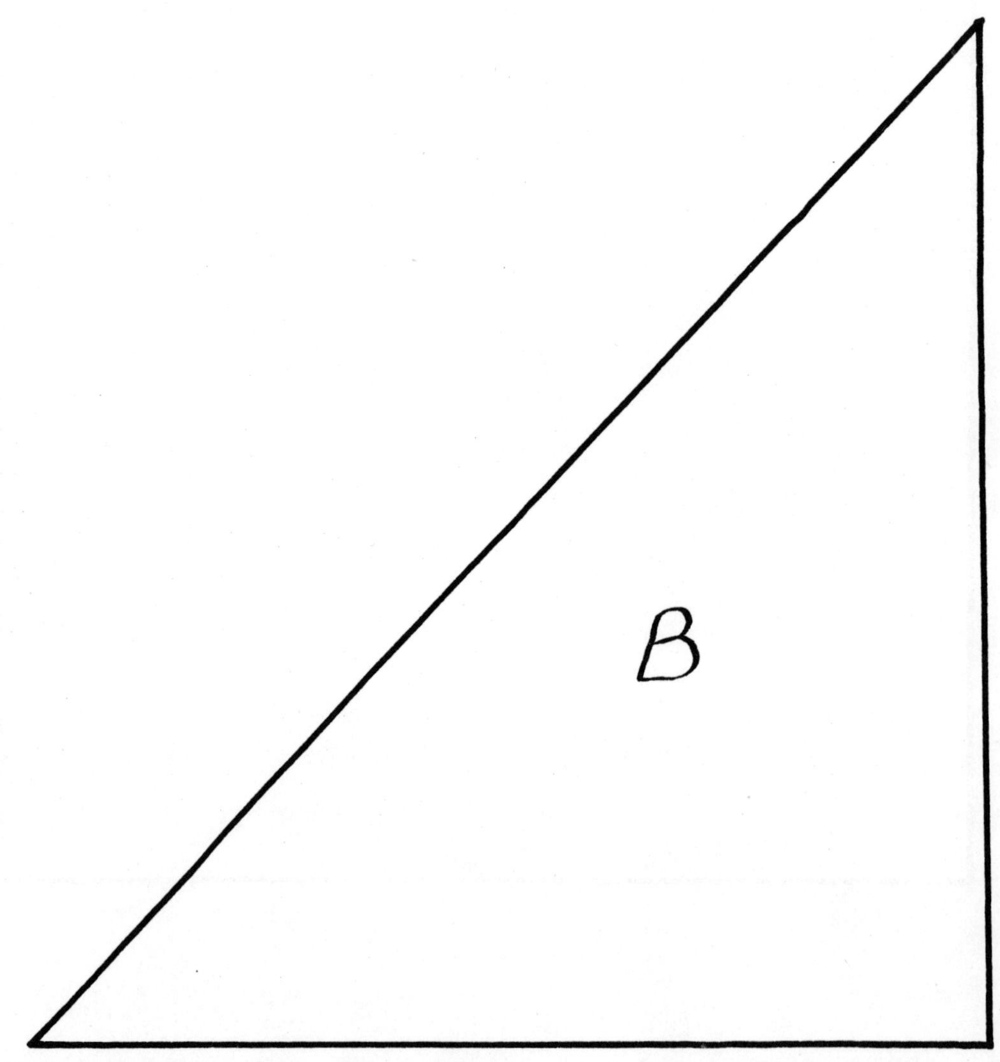

Center Medallion block. Cut 1 light A and 4 dark B. Block finishes approximately 10 inches square.

A

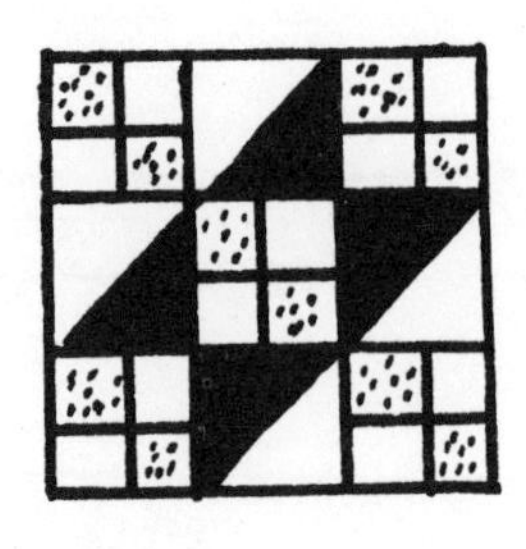

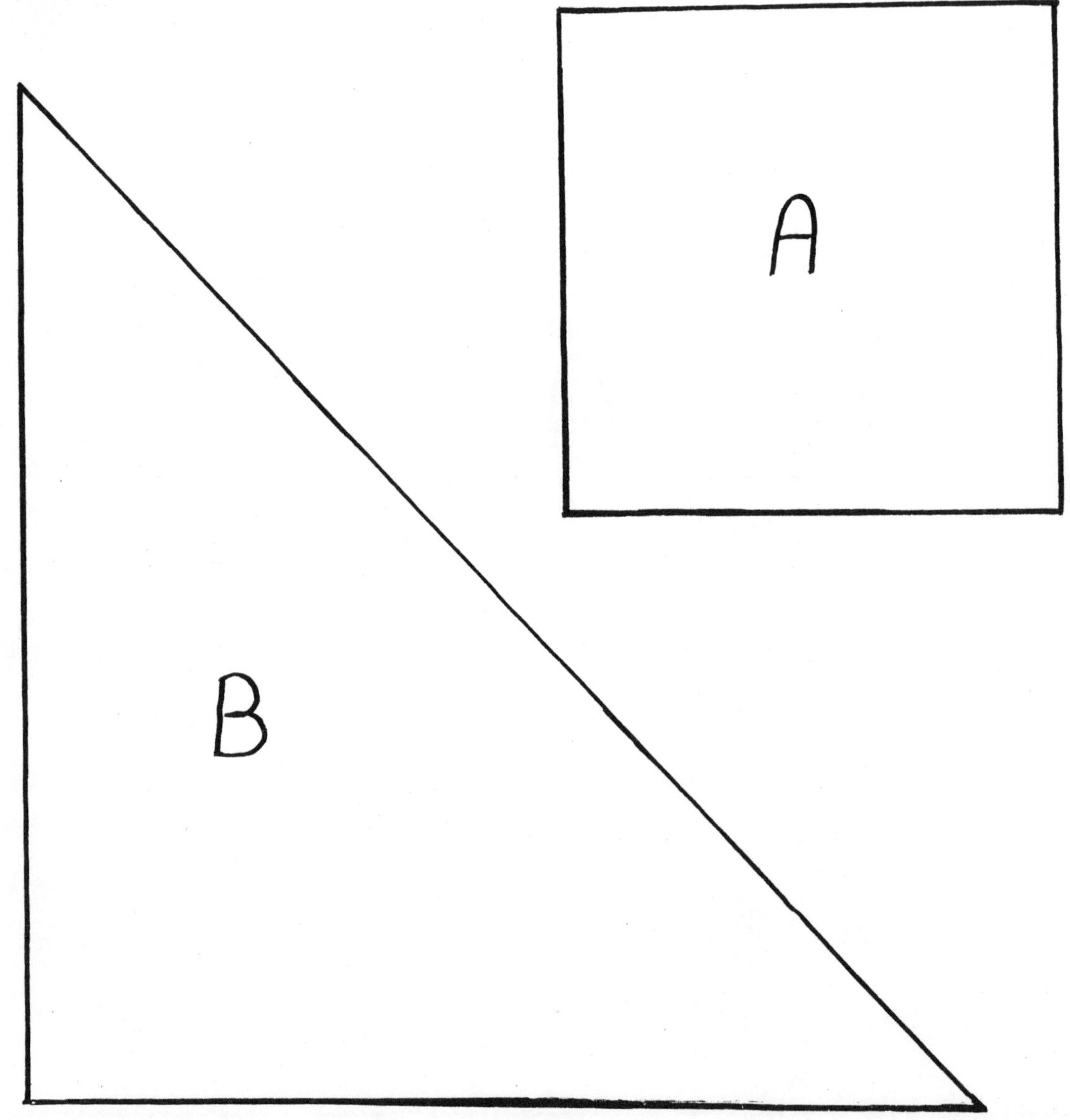

Jacob's Ladder block. Cut 10 light and 10 medium A, cut 4 light and 4 dark B. Block finishes approximately 13 inches square.

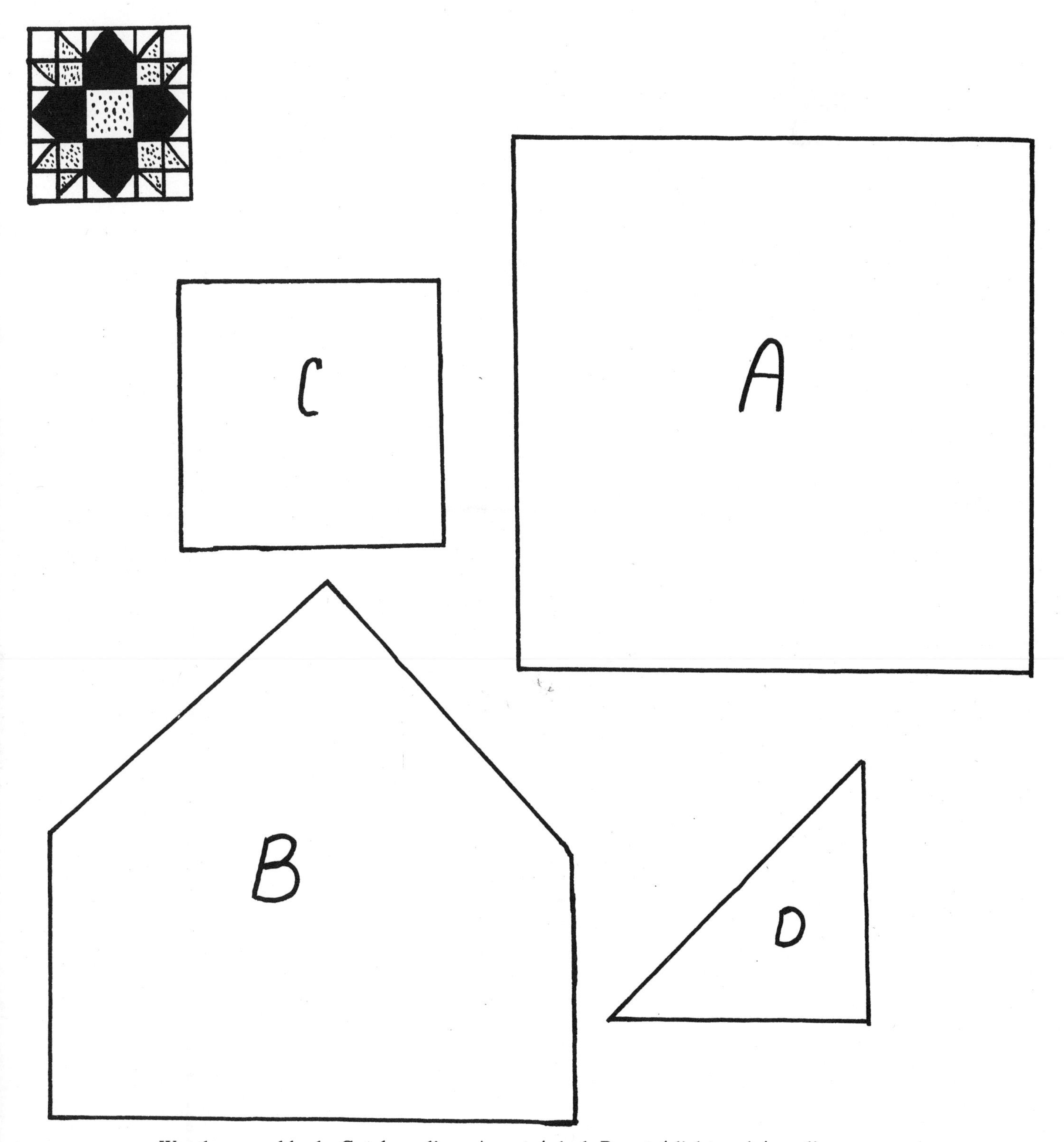

Weathervane block. Cut 1 medium A, cut 4 dark B, cut 4 light and 4 medium C, and cut 8 medium and 16 light D. Block finishes approximately 9 inches square.

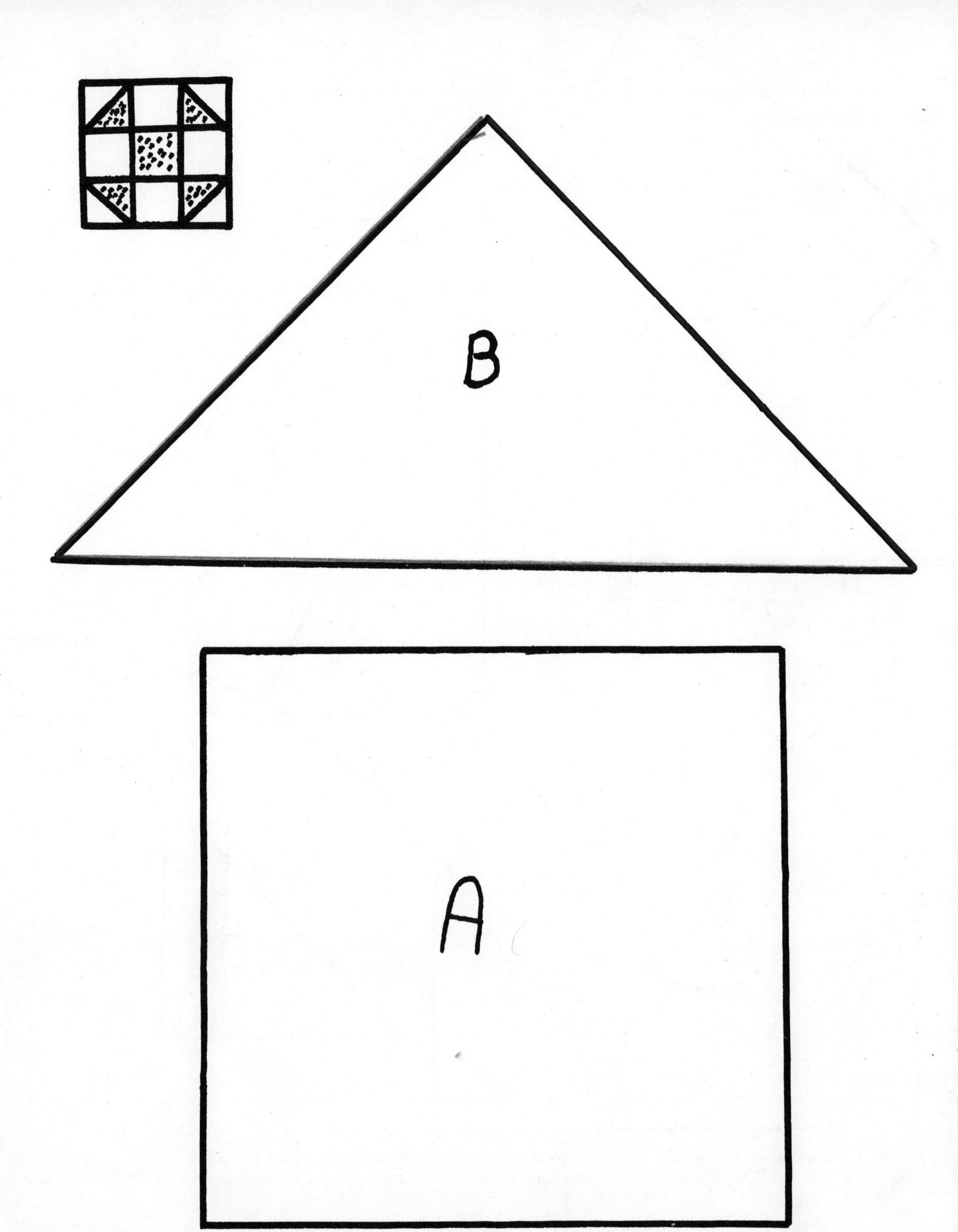

Shoo-Fly block. Cut 1 dark and 4 light A, cut 4 dark and 4 light B. Block finishes approximately 12 inches square.

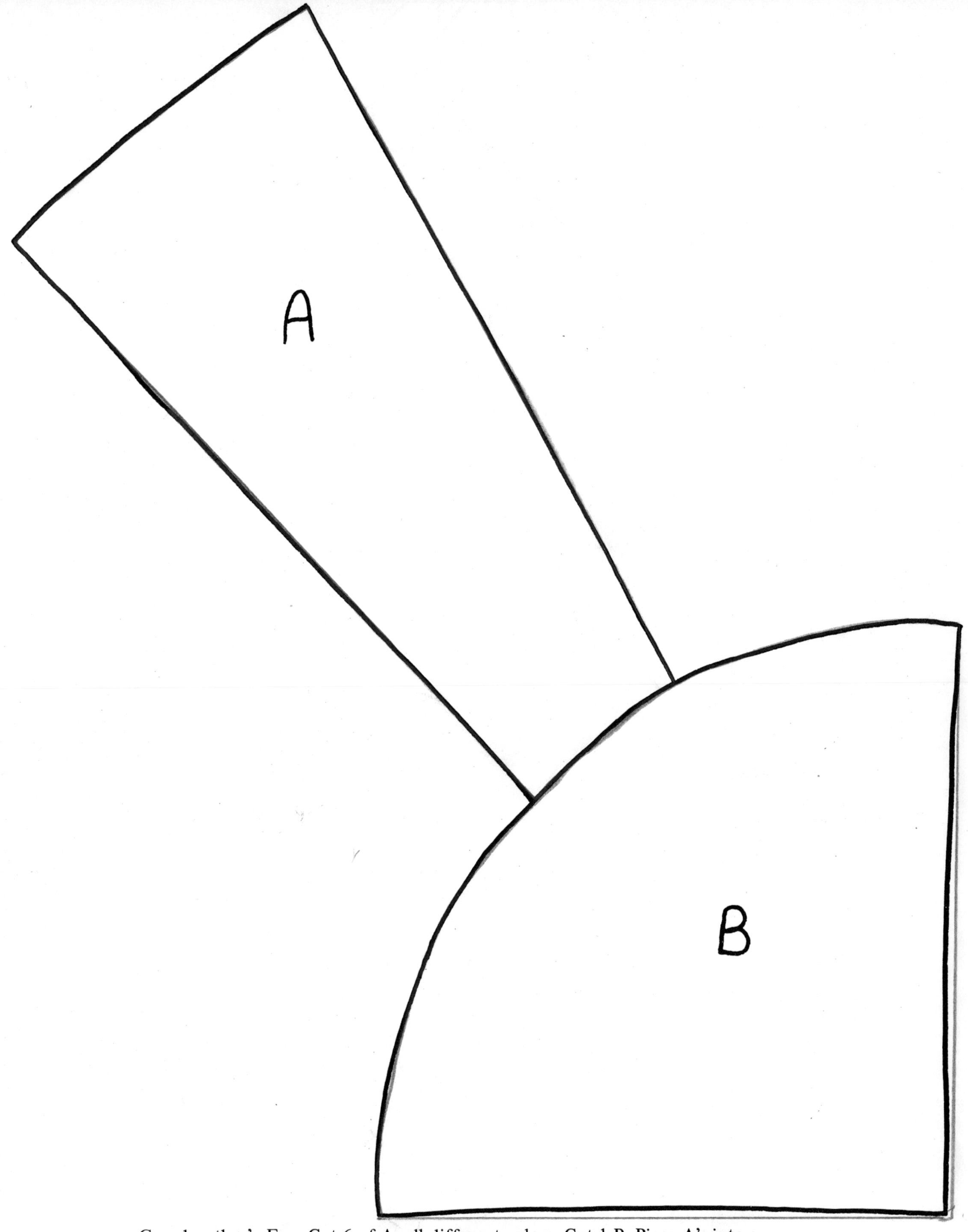

Grandmother's Fan. Cut 6 of A, all different colors. Cut 1 B. Piece A's into a quarter circle, then pin them to a corner of a 13 inch block. Pin B at same corner and appliqué curved edge over bottom raw edges of pieced A's. Then appliqué top raw edge of A's. Finishes 13 inches square.

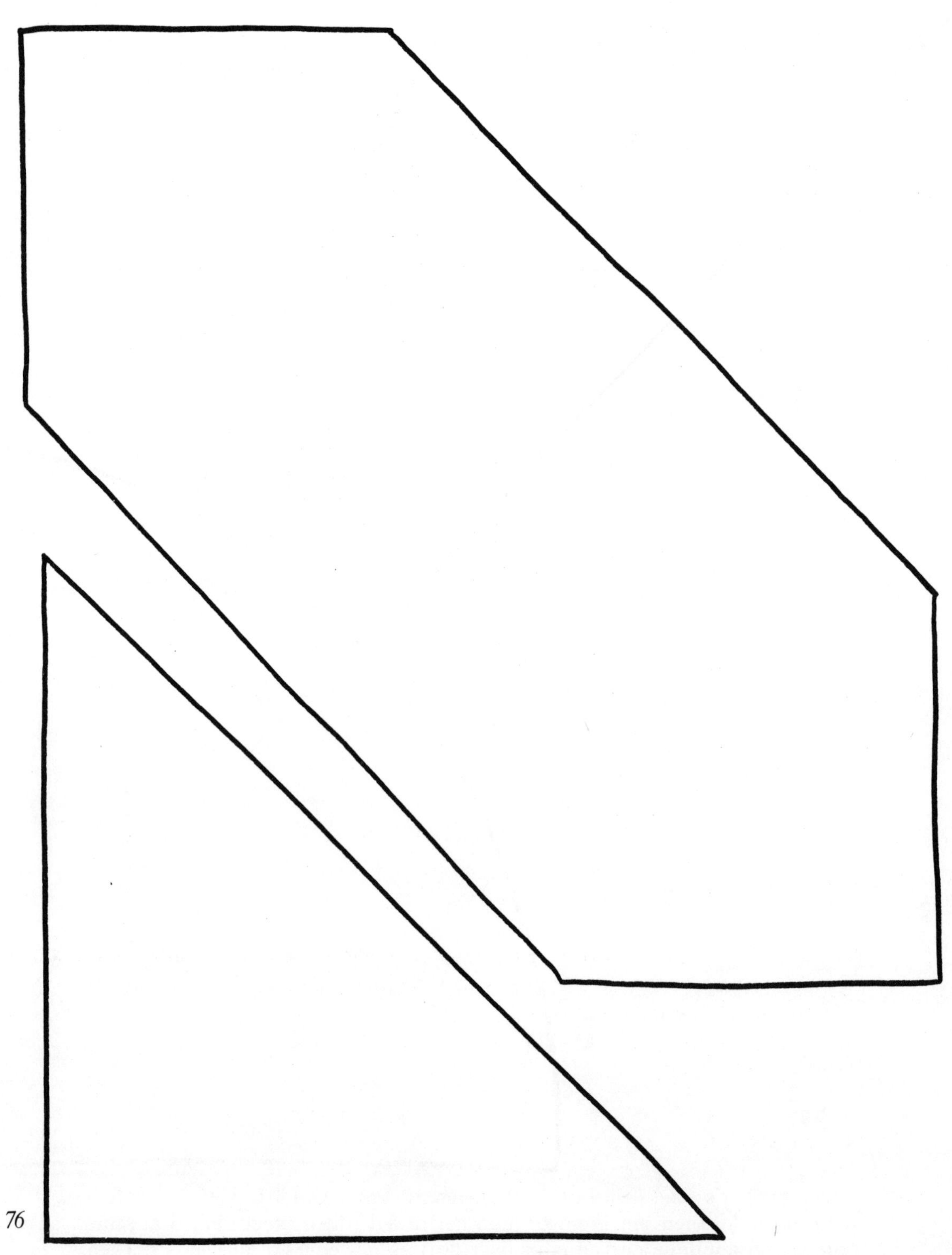

Robbing Peter to Pay Paul. Cut 8 dark triangles. Cut 4 light shapes. Finishes approximately 12 inches square.

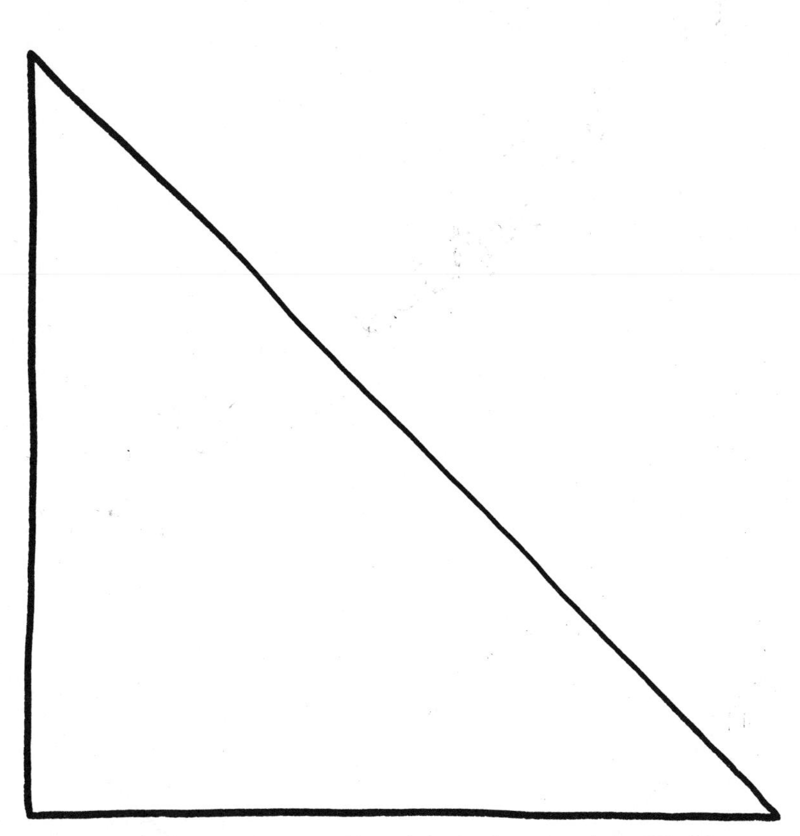

Broken Dishes. Cut 16 light triangles. Cut 16 dark triangles. Finishes approximately 14 inches square.

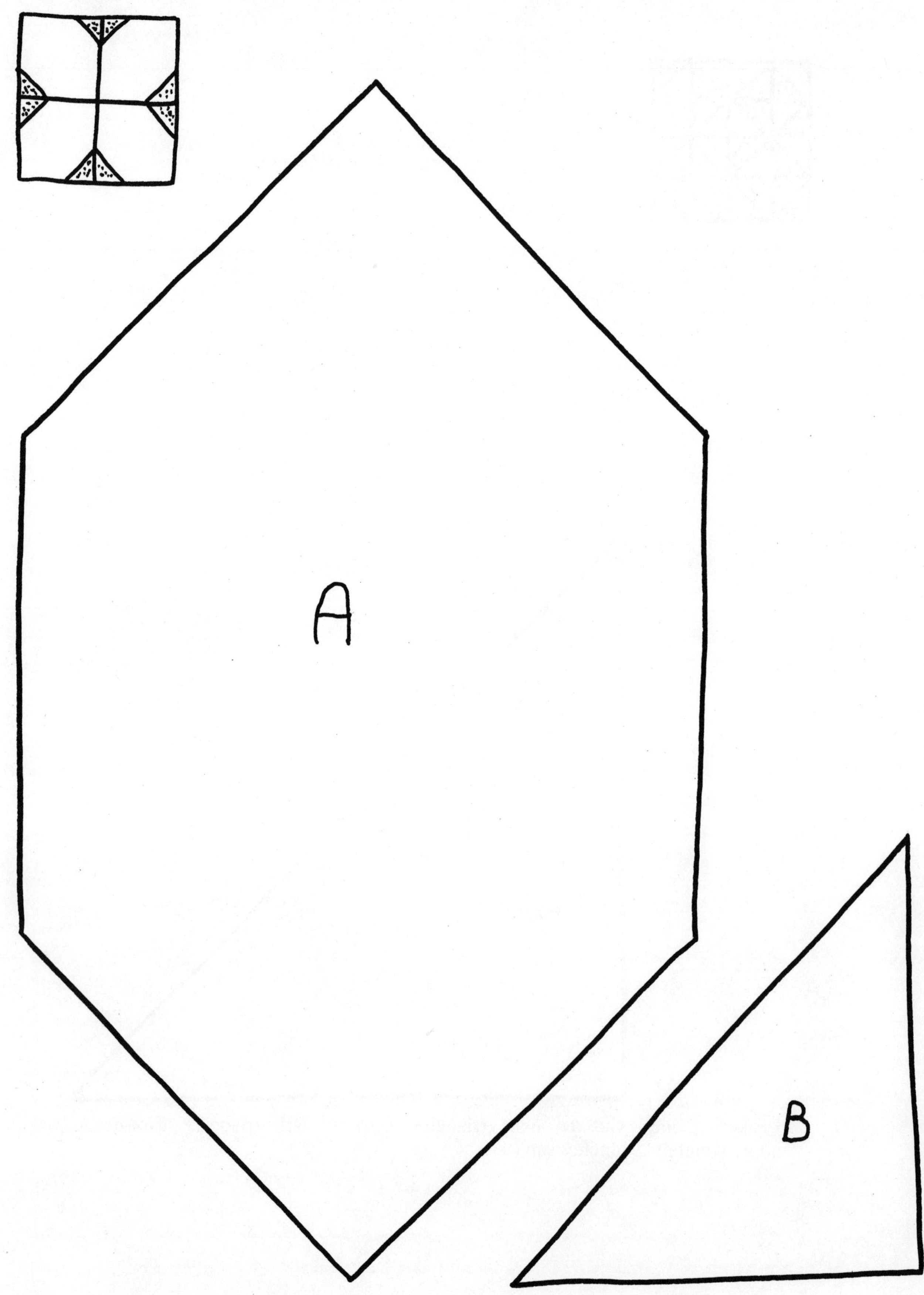

Tomahawk. Cut 4 light A. Cut 8 dark B. Finishes approximately 11 inches square.

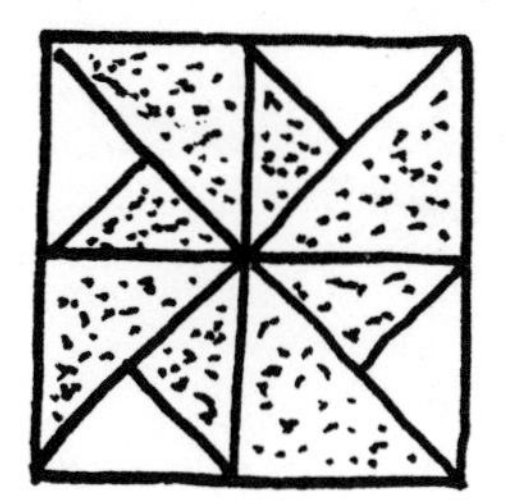

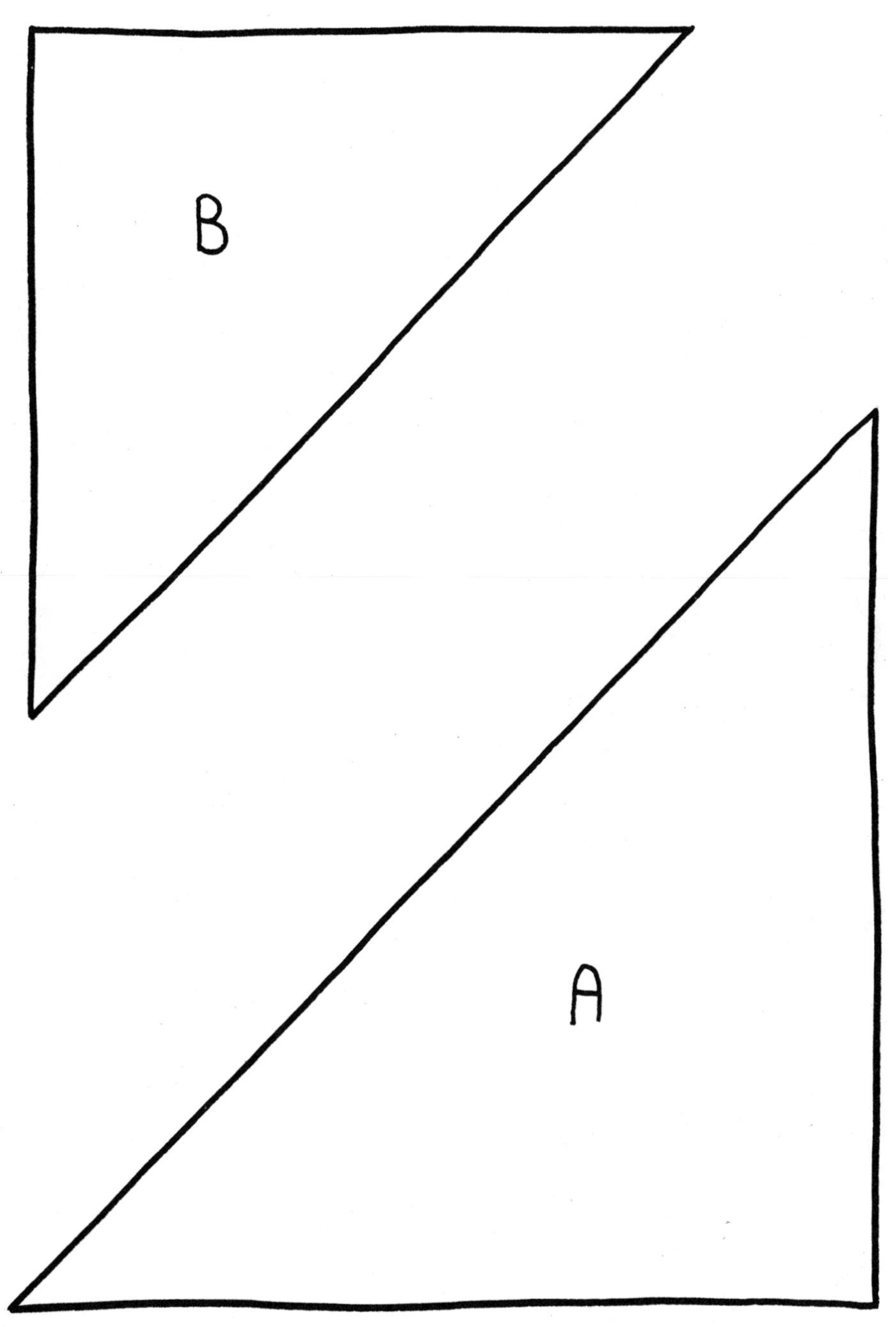

Pinwheel. Cut 4 dark A. Cut 4 dark B. Cut 4 light B. Finishes approximately 8 inches square.

Lemoyne Star or Lemon Star. Cut 8 diamonds, all different. Piece into star shape, then appliqué in center of 15 inch block. Finishes approximately 15 inches square.

Roman Stripe. Cut 4 rectangles of same color. Cut 8 rectangles of different colors. Piece with same color rectangle always in center of each group of three. Finishes approximately 9 inches square.

Detail of Noah's Ark.

Detail of Album quilt.

Balloons appliqué. *Designed and made by Nancy Kane.*

Spring wallhanging made for WNET-TV auction, 1976.

Detail of Spring. *Design courtesy Mark Yoseloff.*

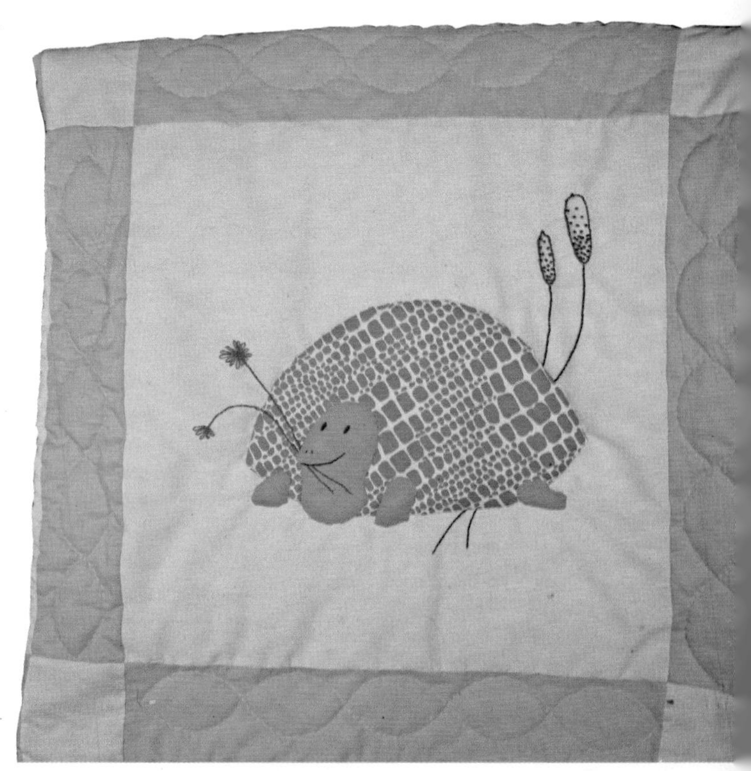

Snowflake, Flower, and Jacob's Ladder pillows. *Made by Geri Bring and Charlotte Healey.*

Lincoln's Platform quilt and pillow and Double Irish Chain pillow. *Made by Marcia Forrest.*

Dresden Plate quilt.

Laura's Quilt. *Designed and made by Geri Bring.*

Weathervane quilt. *Made by Ellie Sommers.*

Star Puzzle quilt.

Hawaiian quilt, modified Pineapple pattern.

Dresden Plate. Cut 18 to 20 petals, all different. Piece edges so that petals form a circle. Eighteen, nineteen, or twenty petals may be used. Appliqué to center of 15 inch block. Finishes approximately 15 inches square.

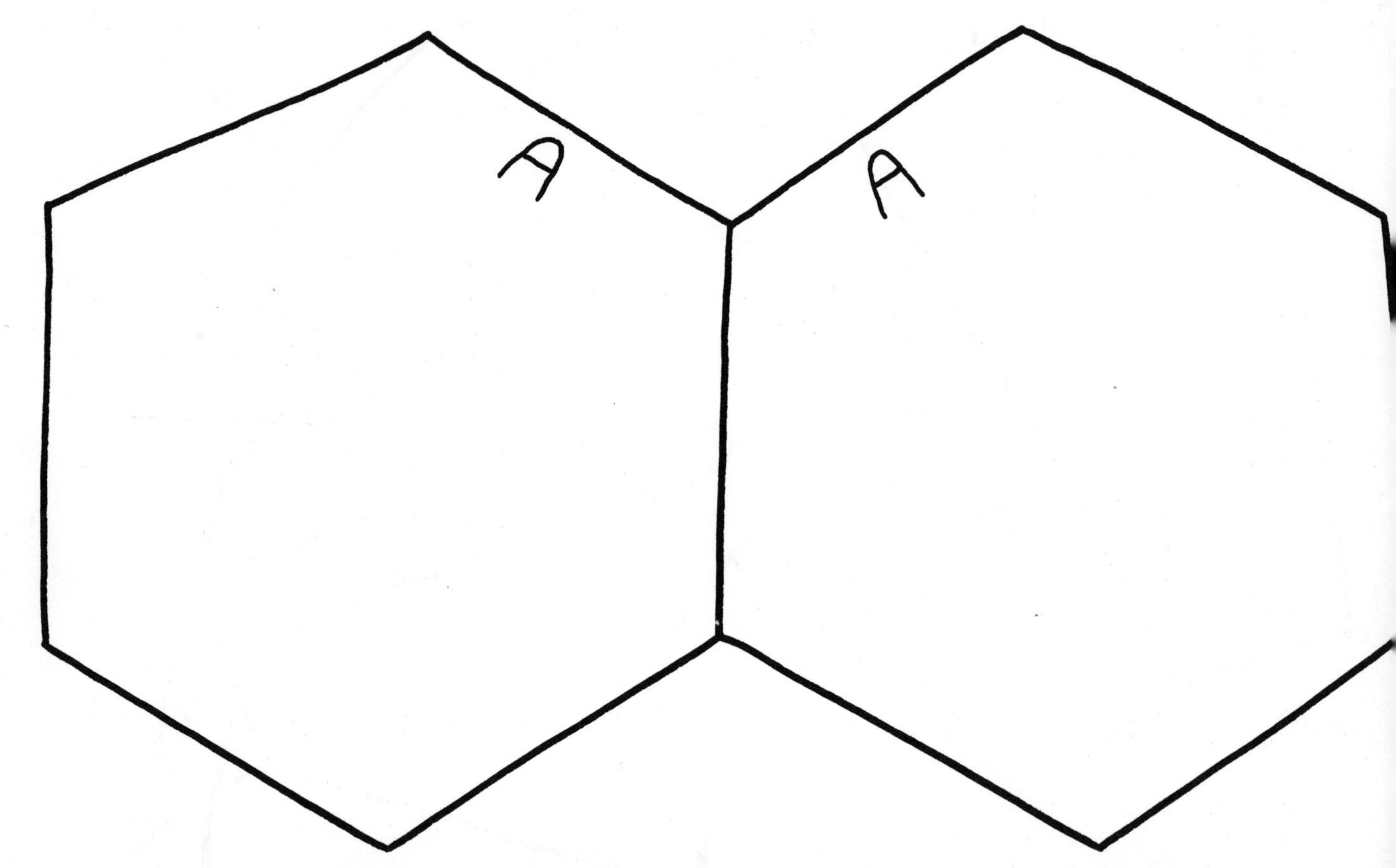

Snowflake. Cut 6 hexagons. Piece edge to edge, forming 6-sided snowflake. Appliqué to center of 12 to 15 inch block.

Balloon appliqué. Cut 7 balloons from various colors. Embroider string in stem stitch.

Daffodil appliqué. Entire design is appliquéd except for stem, embroidered in satin stitch, and dots and legs of ladybug, done in french knots and stem stitch.

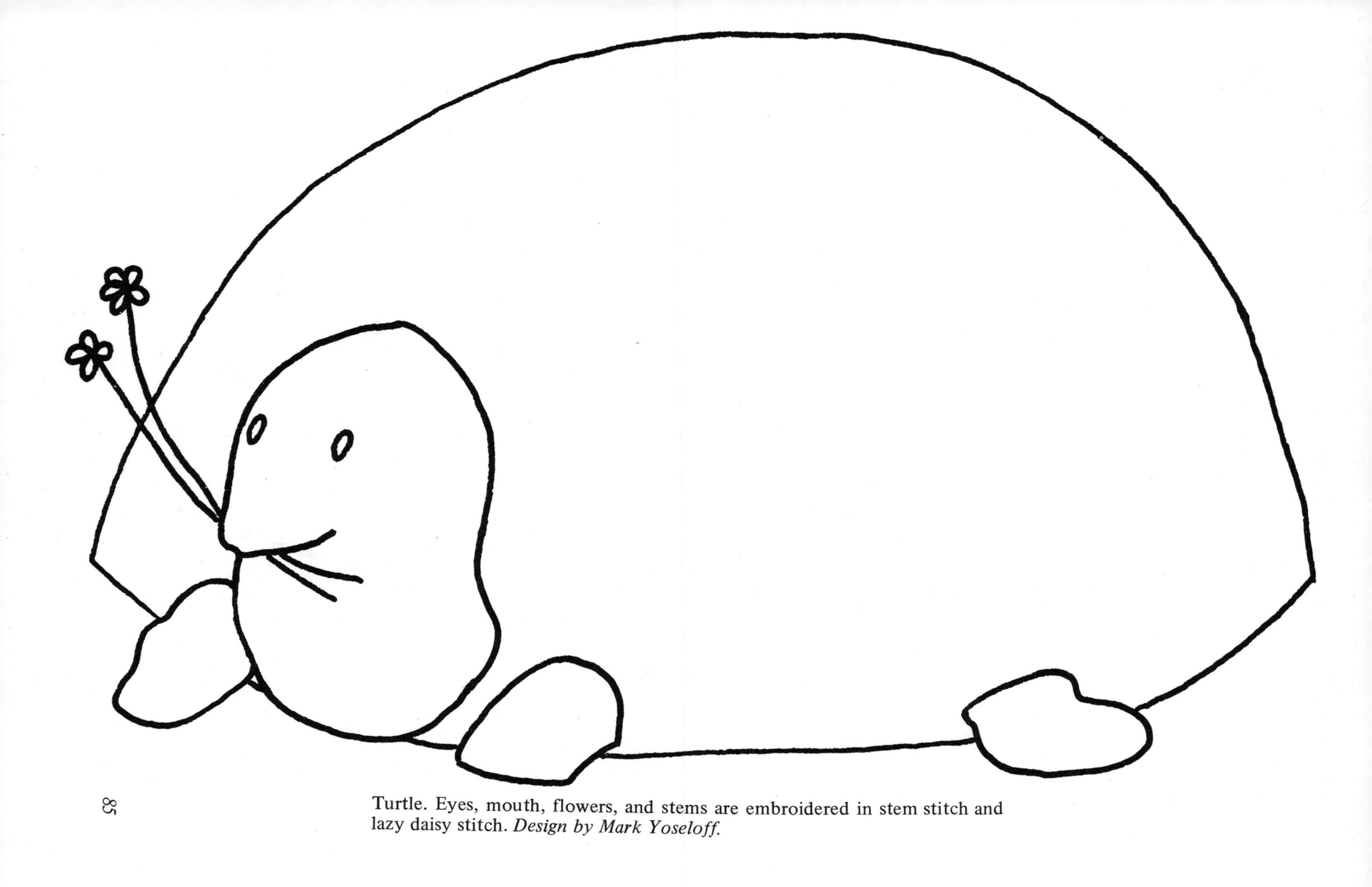

Turtle. Eyes, mouth, flowers, and stems are embroidered in stem stitch and lazy daisy stitch. *Design by Mark Yoseloff.*

Ice Cream Cone. Entire design is appliquéd. Candy sprinkles can be added with french knots.

Snail and Snake. Appliqué with trim of stem stitch and satin stitch.

Frog appliqué. Center of eye, mouth, and leg detail are embroidered.

Mushroom Cluster.

Sunbonnet Baby. Hat, sleeve, and foot are solid color; dress is print. Hand is pink.

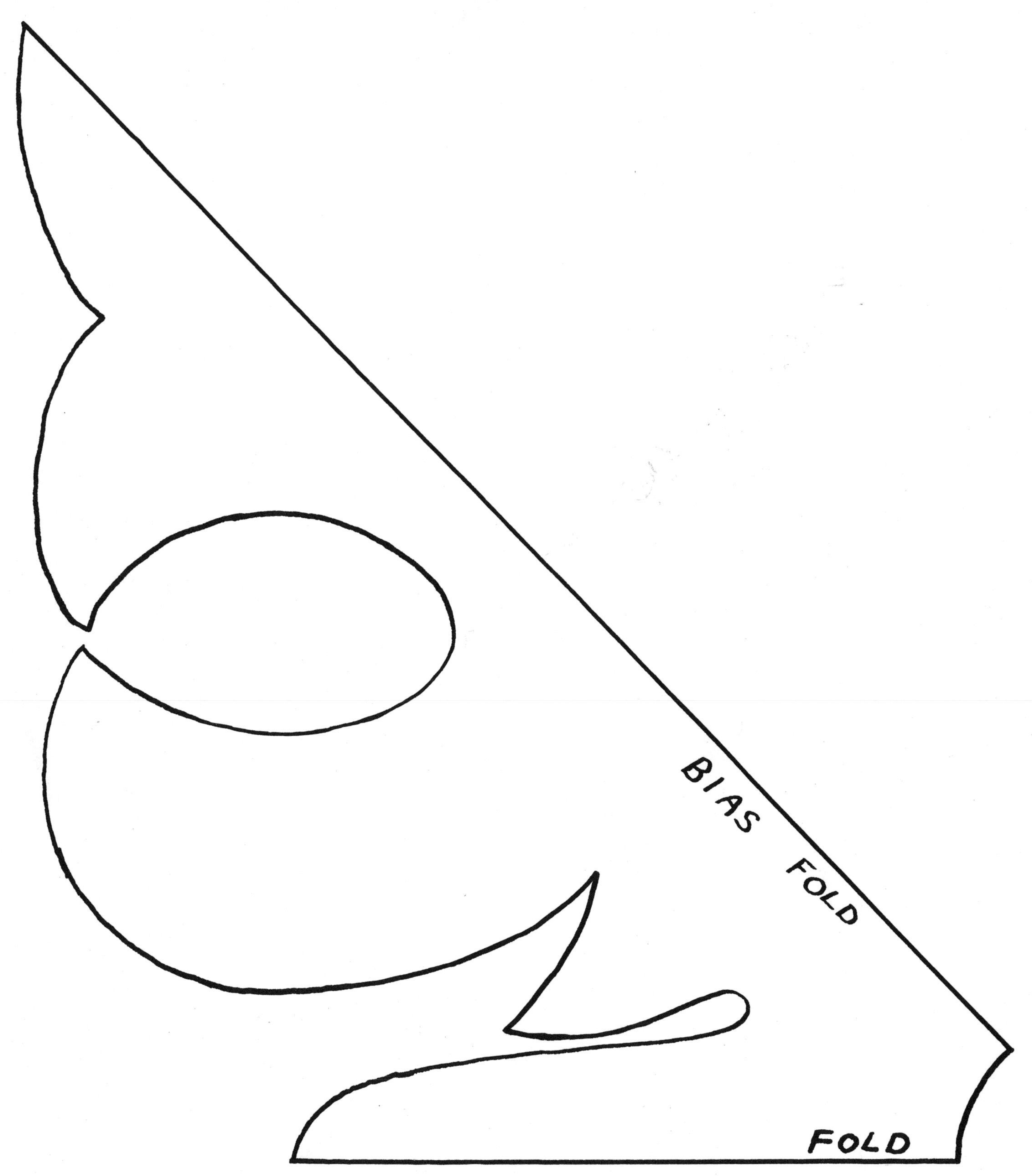

Hawaiian appliqué Flame. Place on a 24-inch background block.

Hawaiian appliqué Hibiscus. Center on a 24-inch background block.

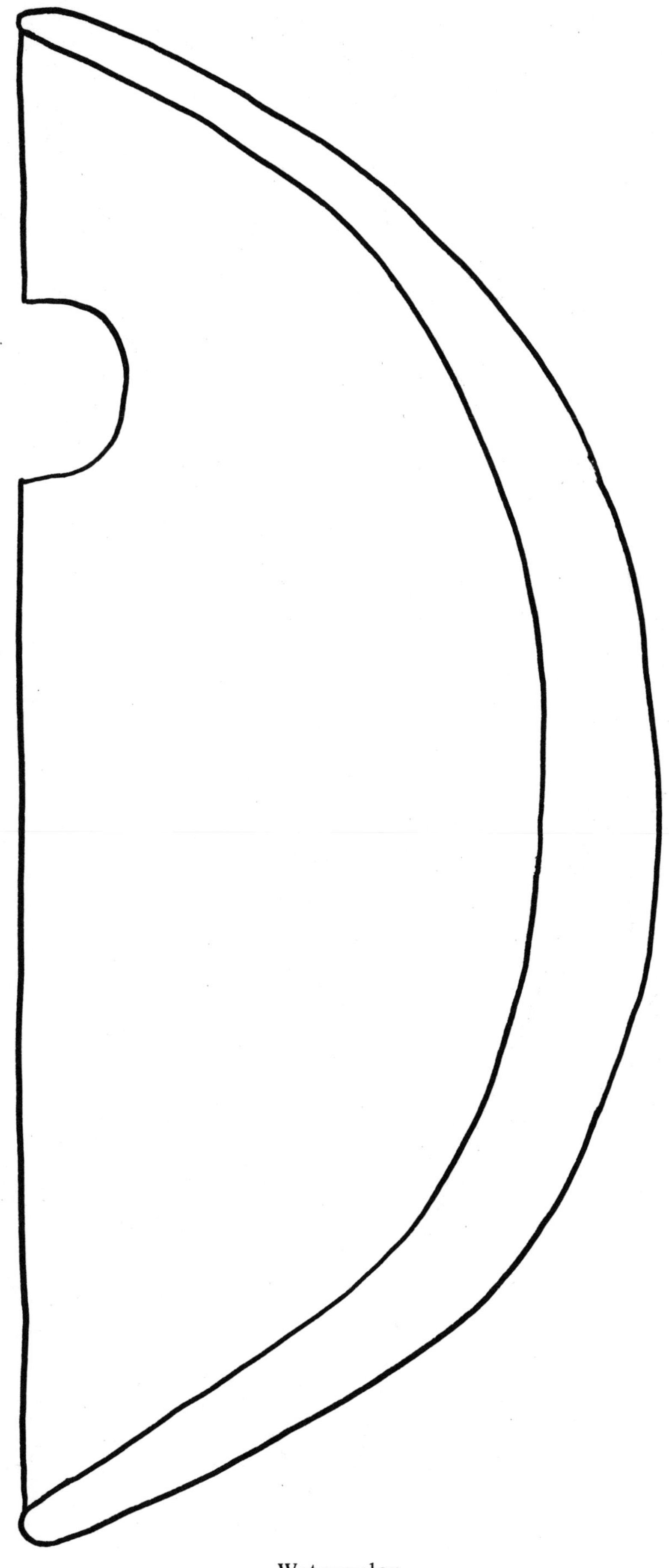

Watermelon.

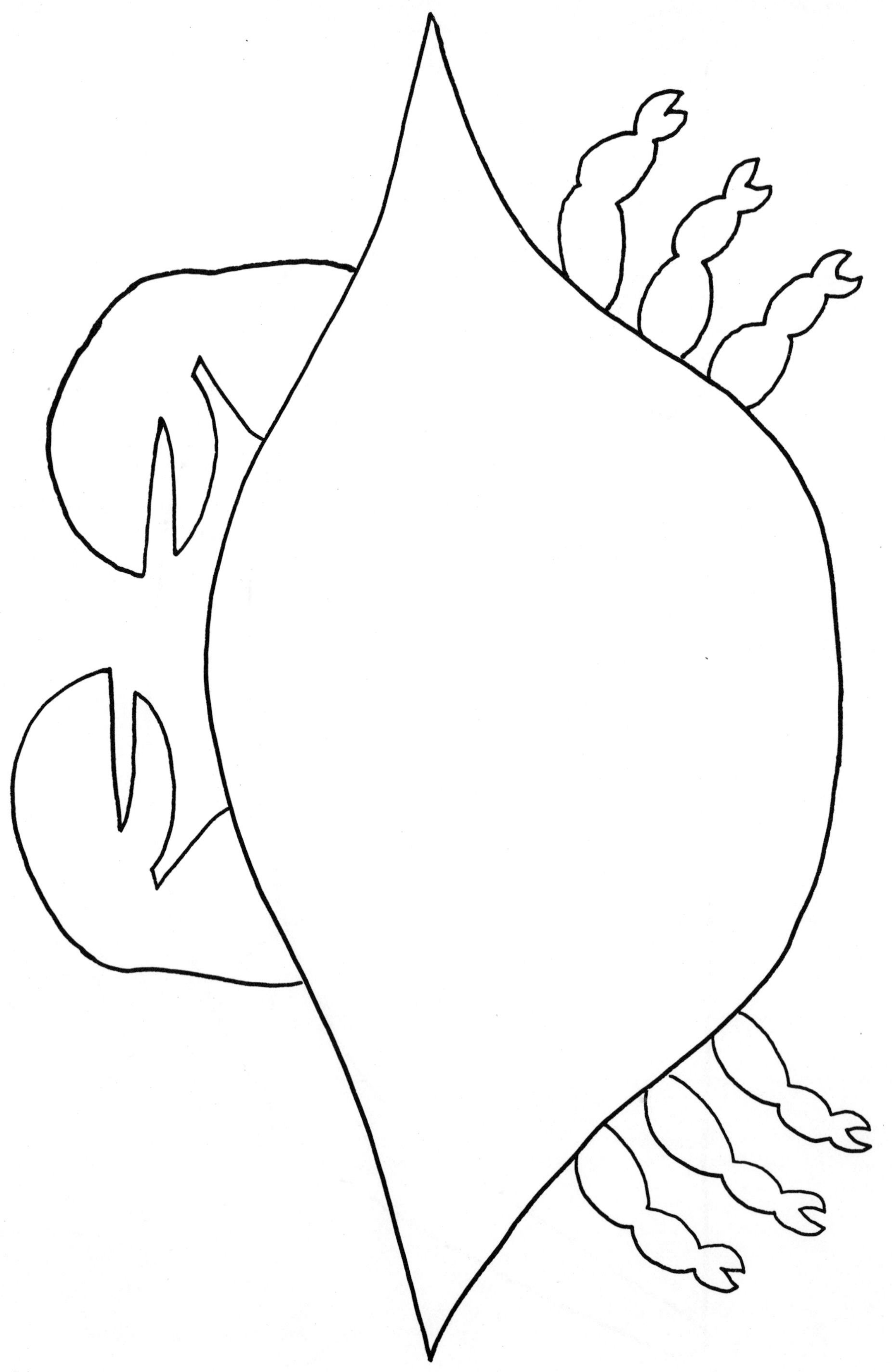

Crab. Appliqué Body and claws. Back flippers are embroidered in outline stitch.

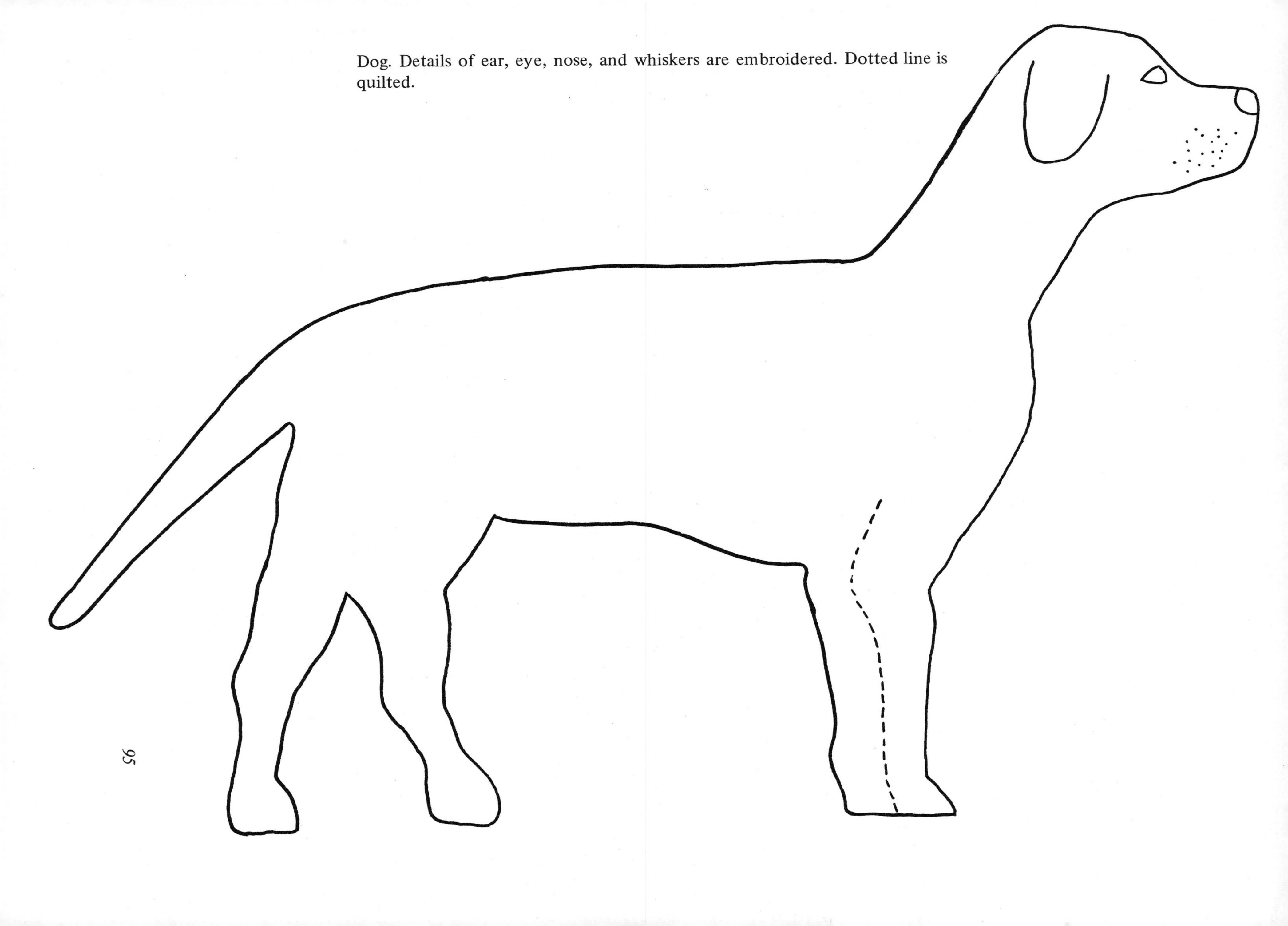

Dog. Details of ear, eye, nose, and whiskers are embroidered. Dotted line is quilted.

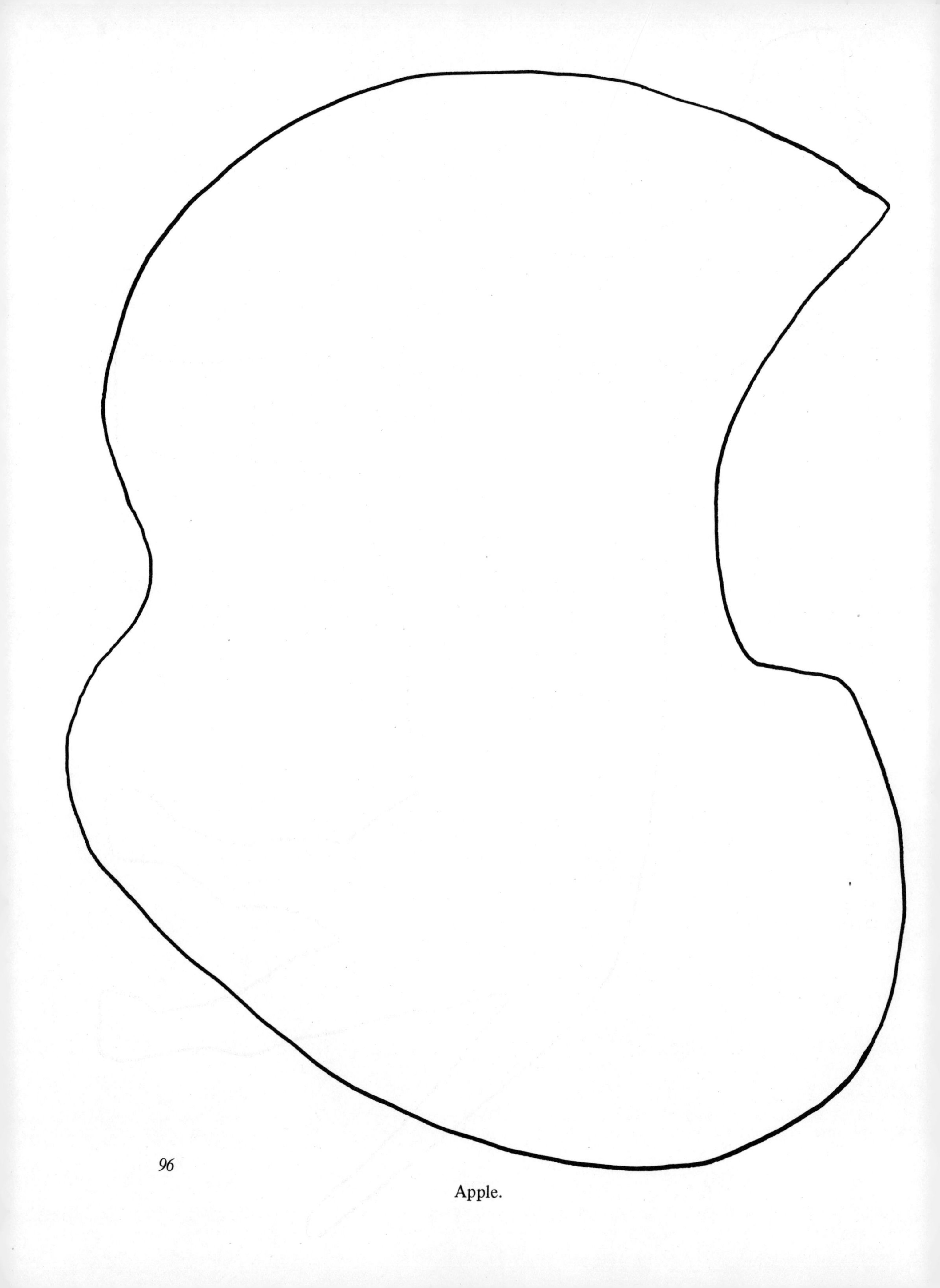

Apple.

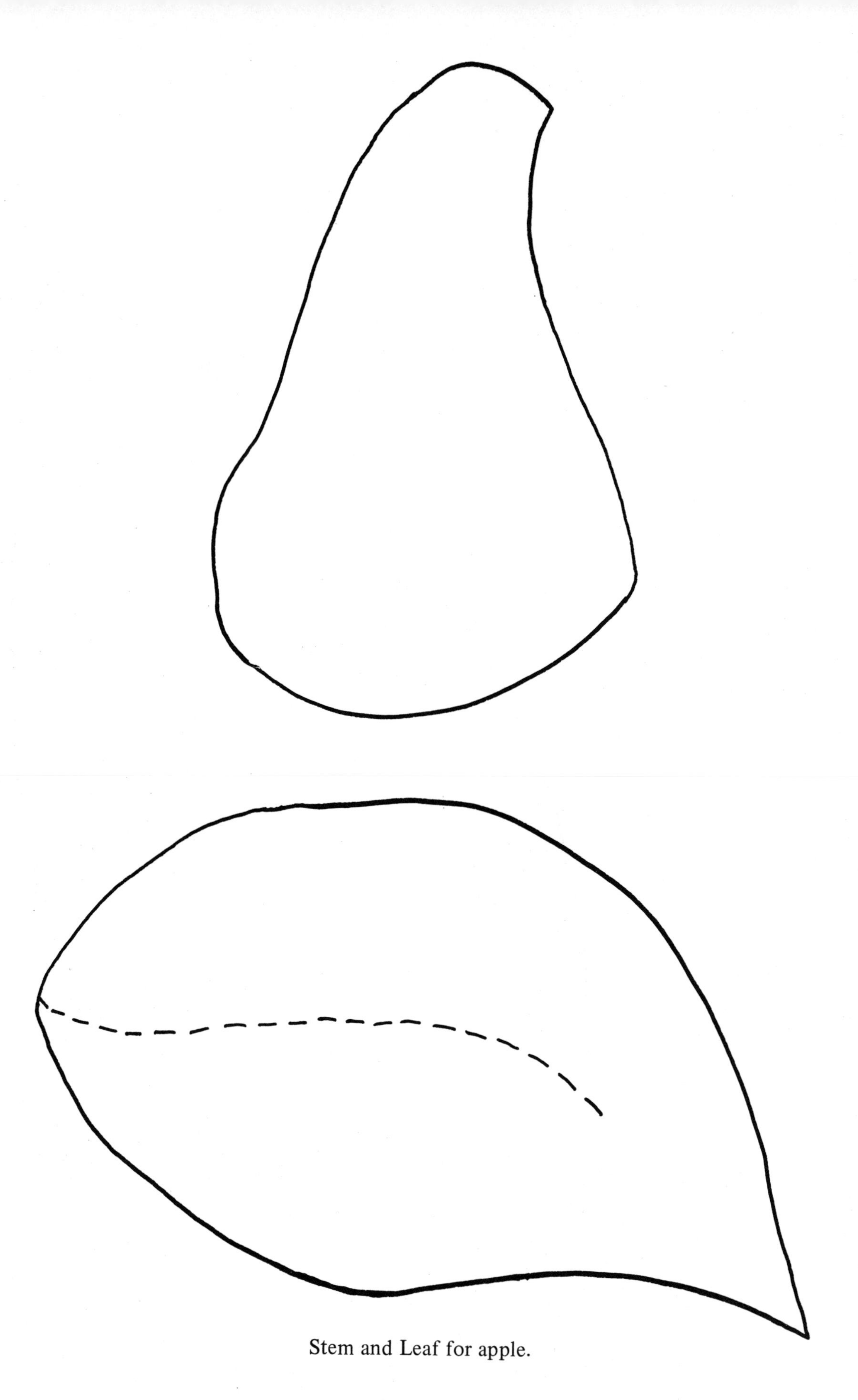

Stem and Leaf for apple.

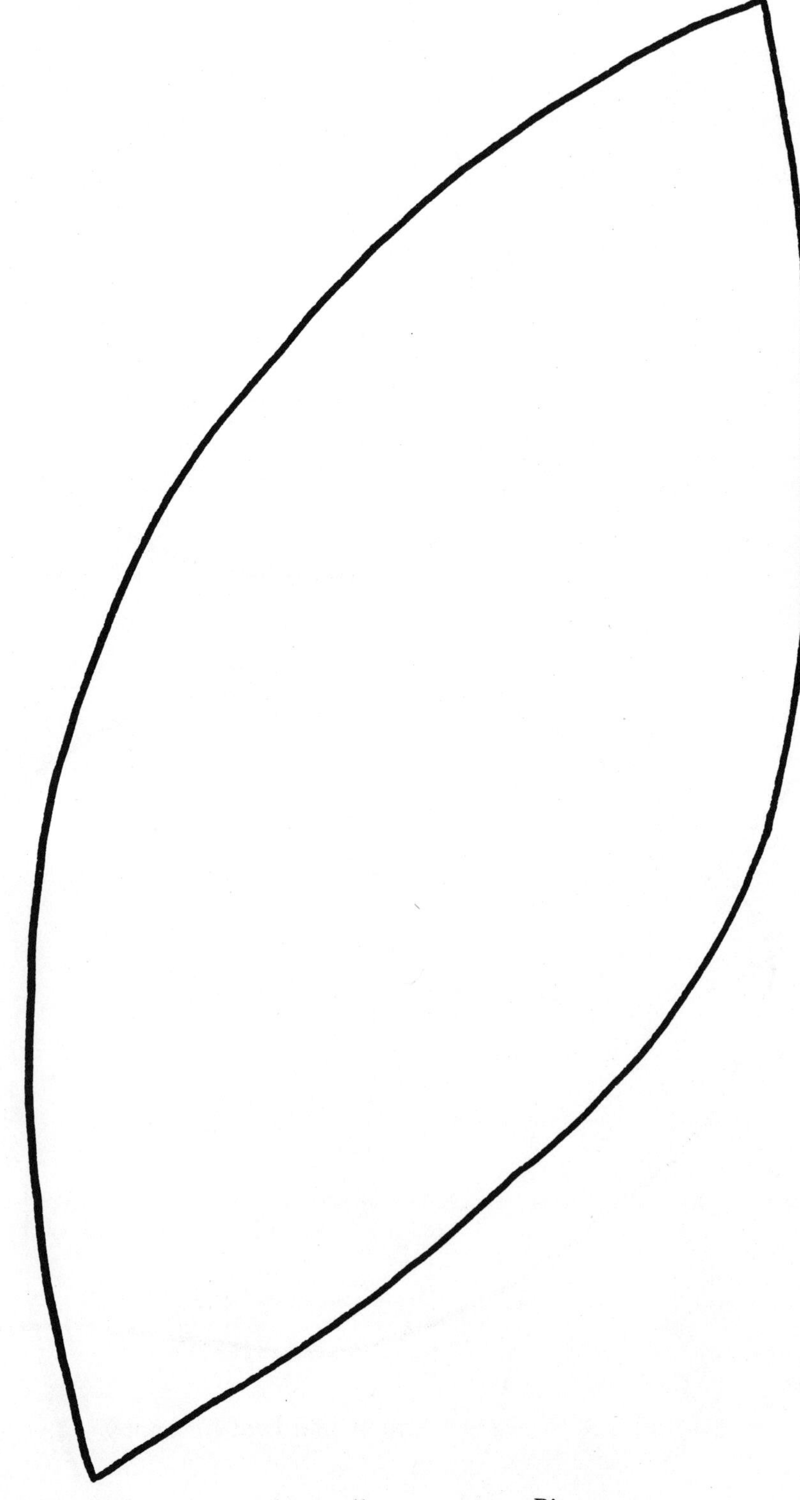

98 Orange Peel. Cut 4 shapes and appliqué diagonally on squares. Piece squares together to form pattern.

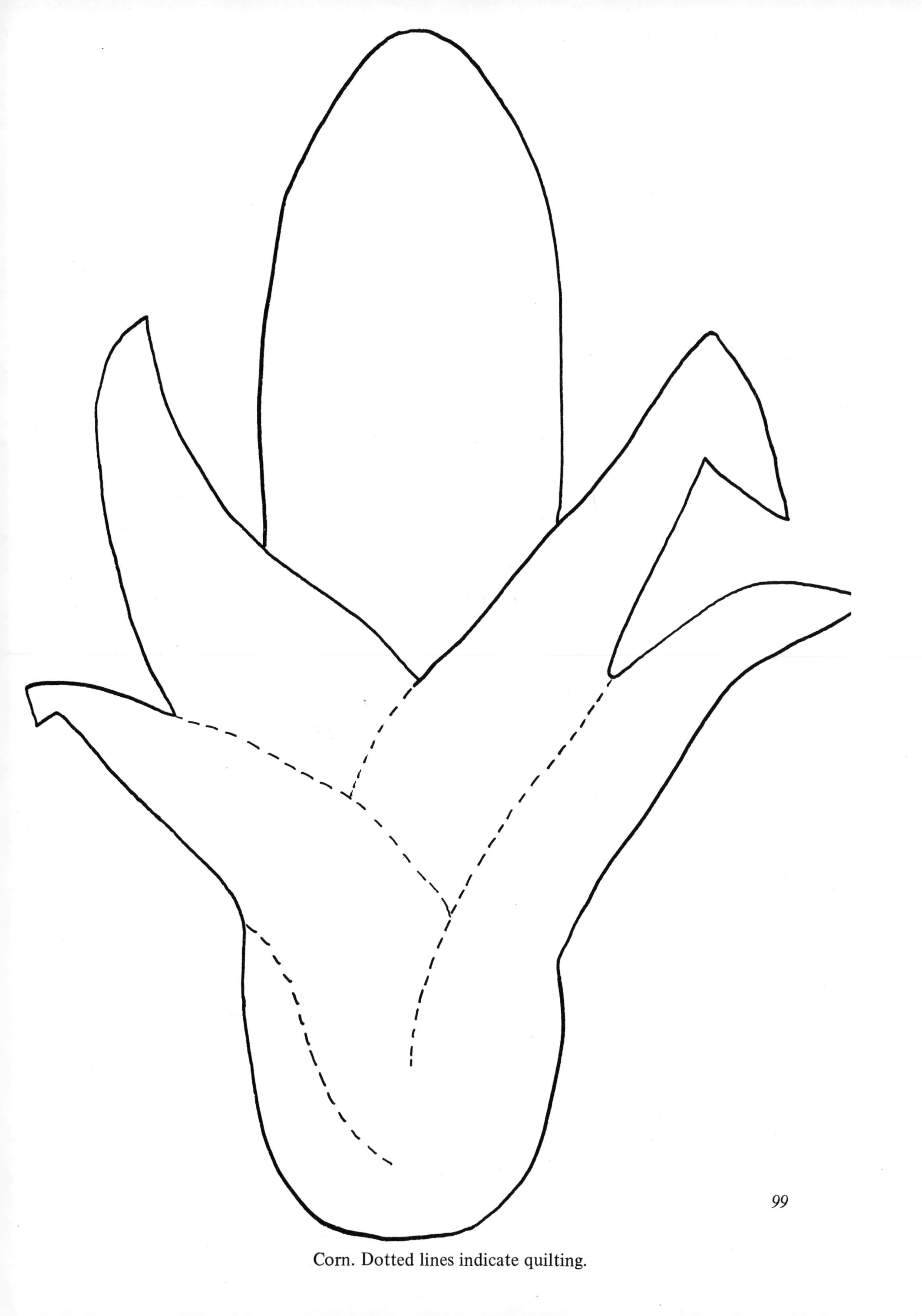

Corn. Dotted lines indicate quilting.

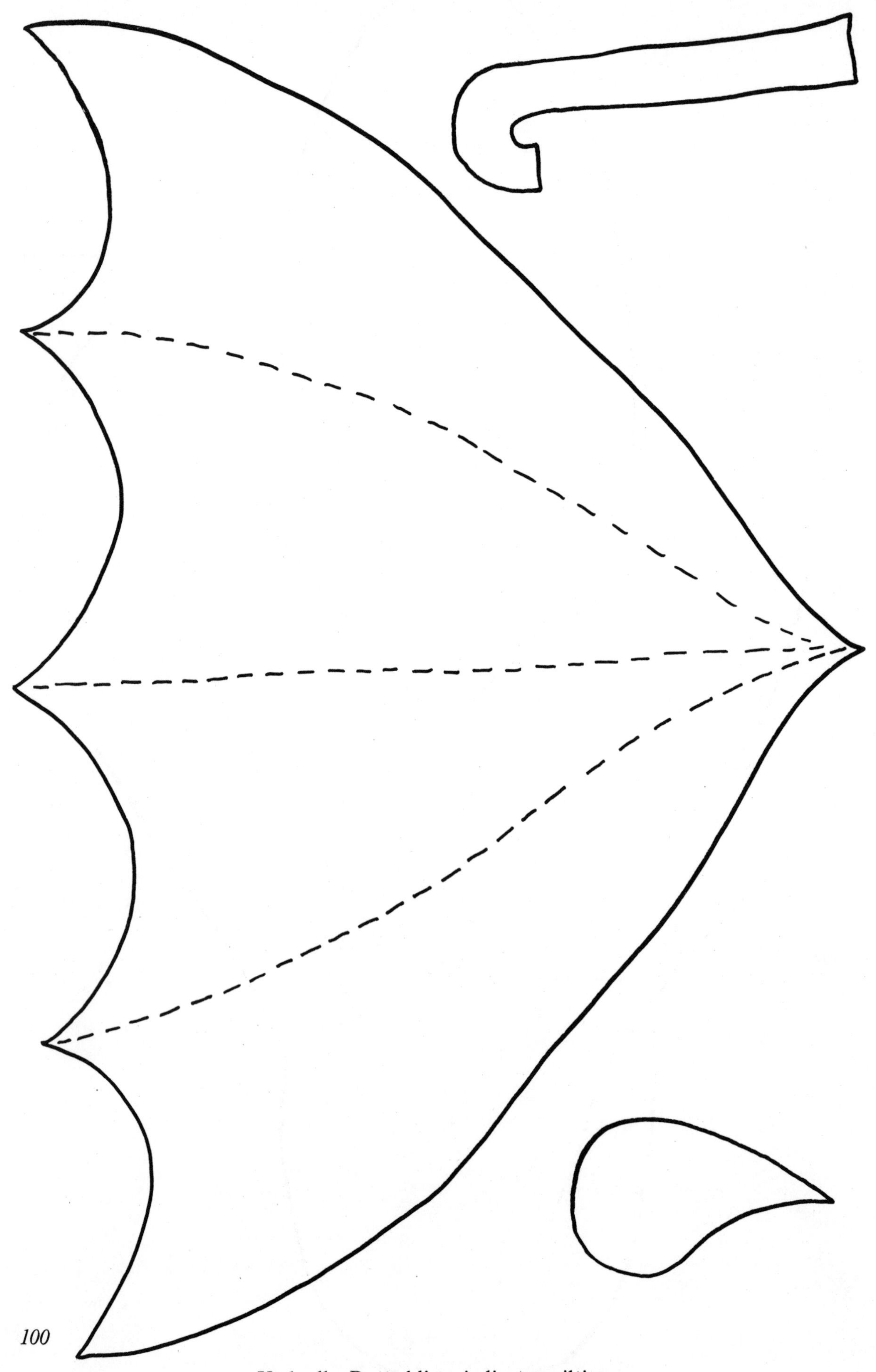

Umbrella. Dotted lines indicate quilting.

Ladybug. Antennas may be embroidered in outline stitch.

Heart.

Fish. Body, eye and tongue are appliqued. Eyeball is embroidered in satin stitch. Fins, mouth, and dream are embroidered in outline stitch.

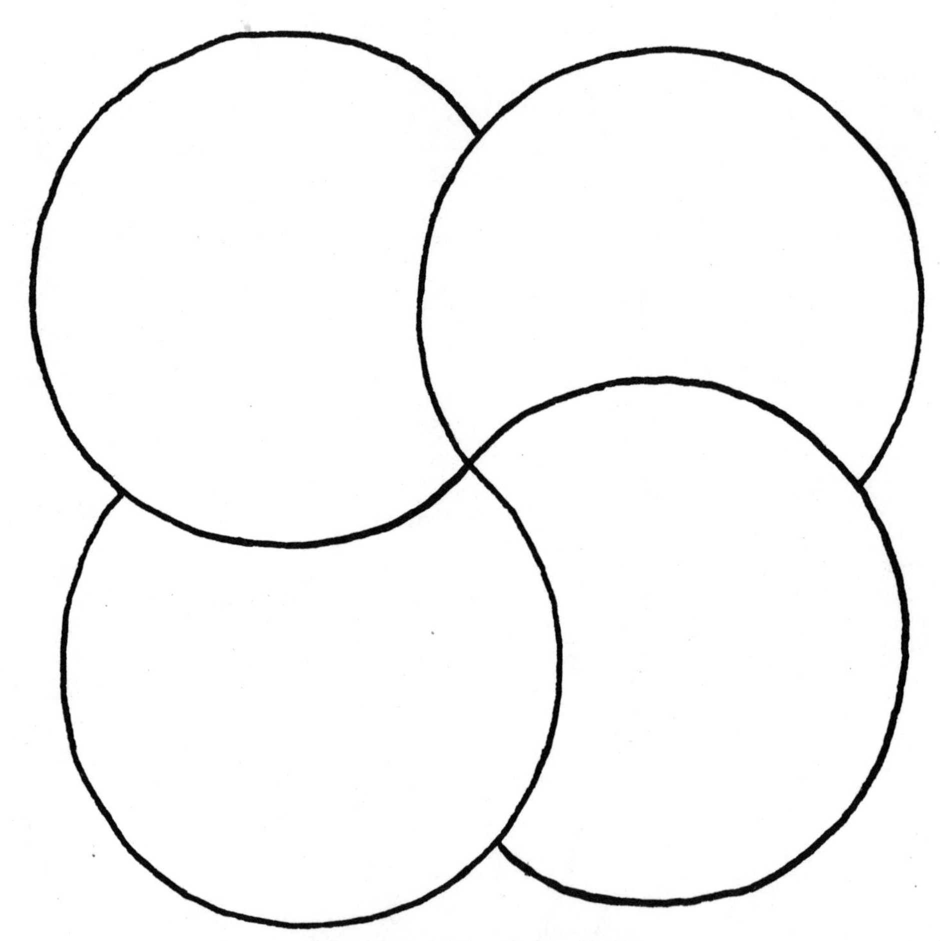

White-on-white design Swirl.

White-on-white design Daisy.

White-on-white designs Small Daisy and Dahlia.

White-on-white designs Leaf and Hearts.

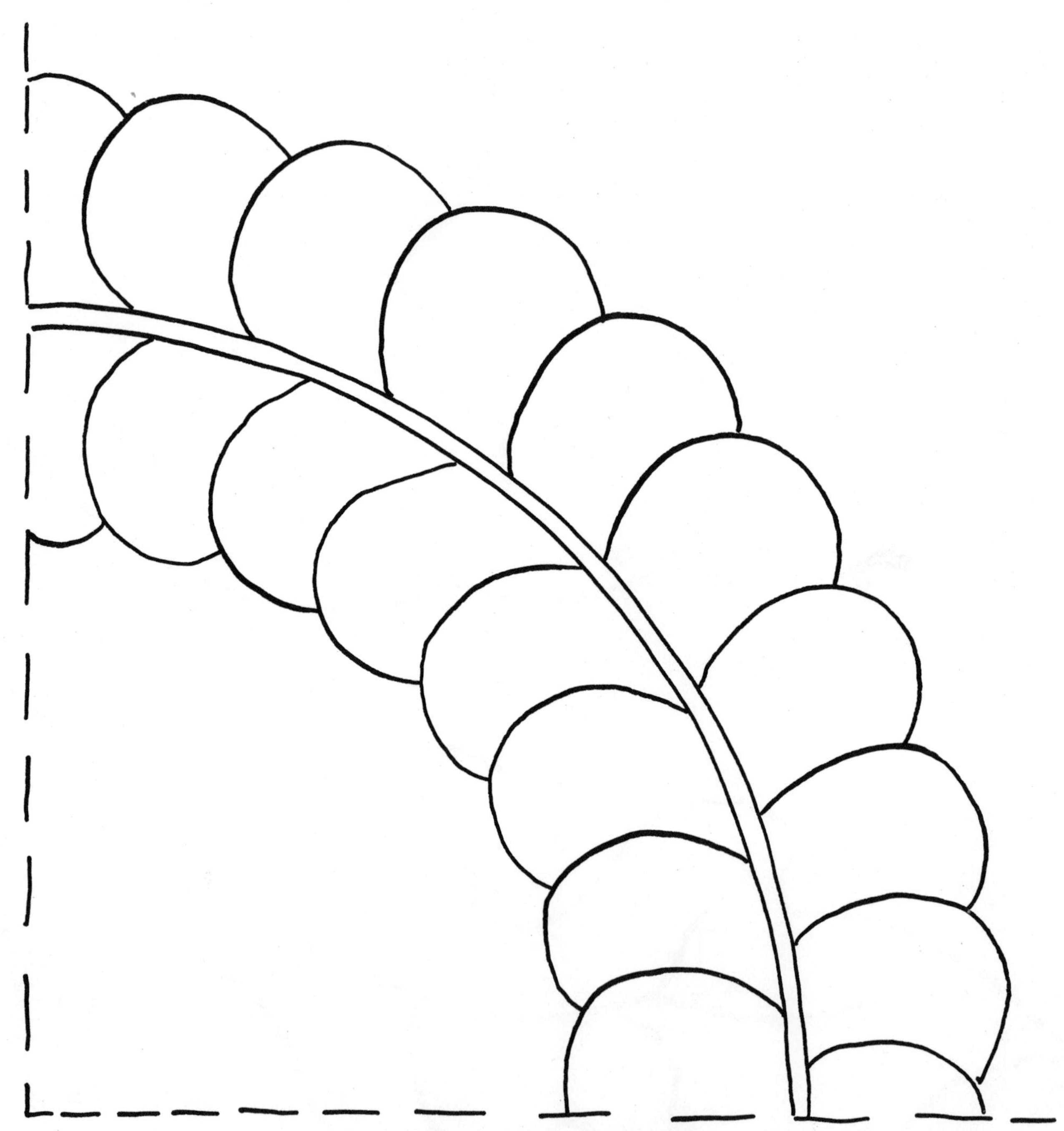

White-on-white design Wreath. Design may be used as quarter wreath as shown, or continued to form half wreath or complete wreath.

White-on-white design Locust Leaf Spray.

Bibliography

Brightbill, Dorothy. *Quilting as a Hobby.* New York: Bonanza Books, 1963. 96 pp.

Gammell, Alice I. *Polly Prindle's Book of American Patchwork Quilts.* New York: Grosset & Dunlap, 1973. 238 pp. The patterns (50 of them) make this book worthwhile. It contains everything from very simple ones to real challenges. Well worth buying.

Hall, Carrie A., and Kretsinger, Rose. *The Romance of the Patchwork Quilt in America.* Caldwell, Idaho: Caxton, 1935. 299 pp. An inexpensive reprint is available. This is the book I use most in identifying patterns.

Hinson, Dolores A. *A Quilter's Companion.* New York: Arco, 1973. 281 pp.

Holstein, Jonathan. *The Pieced Quilt – An American Design Tradition.* Greenwich, Conn.: N.Y. Graphic Society, 1973. 192 pp. The text is over my head, but the pictures are magnificent. An inspiration!

Houk, Carter, and Miller, Myron. *American Quilts and How To Make Them.* New York: Scribners, 1975. 200 pp.

Ickis, Marguerite. *The Standard Book of Quilt-making and Collecting.* New York: Greystone Press, 1949. 273 pp. I have the Dover reprint. Considered a standard work in the field.

Lane, Rose Wilder. *Woman's Day Book of American Needlework.* New York: Simon & Schuster, 1963. 208 pp. Magnificent photographs, fascinating text. There are good sections on patchwork, appliqué, and white-on-white quilting.

Laury, Jean Ray. *Quilts and Coverlets – A Contemporary Approach.* New York: Van Nostrand Reinhold, 1970. 128 pp. Full of inspiration.

Leman, Bonnie. *Quick and Easy Quilting.* New York: Hearthside, 1972. 191 pp. Excellent instructions for many types of no-frame quilting.

Lewis, Alfred Allan. *The Mountain Artisan's Quilting Book.* New York: Macmillan, 1973. 179 pp.

McKain, Sharon. *The Great Noank Quilt Factory.* New York: Random House/Pequot Press, 1974. 136 pp.

McKim, Ruby Short. *One Hundred and One Patchwork Patterns.* New York: Dover, 1963. 124 pp. Paperback. Beautiful, traditional patterns.

Marein, Shirley. *Stitchery, Needlepoint, Appliqué and Patchwork.* New York: Viking, 1974. 207 pp.

Pforr, Effie Chalmers. *Award Winning Quilts.* Birmingham, Ala.: Oxmoor House, 1974. 184 pp. (Progressive Farmer Book). Challenging patterns with each block pictured in color.

Safford, Carleton L., and Bishop, Robert. *America's Quilts and Coverlets.* New York: Dutton, 1972. 313 pp.

Sunset Books editors. *Quilting and Patchwork.* Menlo Park, Calif.: Lane Books, 1973. 80 pp. Paperback.

Wooster, Ann-Sargent. *Quiltmaking.* New York: Drake, 1972. 166 pp.

Index

References to illustrations are in boldface type.

Album quilt, 27
Apple, **96**, **97**
Alternating block design, 35

Backing fabric, 28, 29, 38, 39, 44
Balloons, **83**
Basketweave, **59**
Basting, 27, 28, **28**, 29, 44, **45**, 51
Batt, 37, 38, 39, 43, 44, 57; cotton, 11, 28, 57, 58, 59; flannel, 28, 57; piecing, 29; polyester, 28, 29; trims, 51; wool, 11, 57
Bearpaw, 11, **13**
Bedcoverings, European, 11
Bedspreads, 46
Bench pads, 56
Bias binding, 40
Bicentennial, 12
Binding, replacement of, 57; stuffed, 43
Biscuit, 53, **53**, 54, **54**, 55, **55**, 56, **62**
Blanket stitch, 23, **23**
Blind stitch, 22, **23**, **24**, 25, **25**, 26, **26**, 27, **27**, 38, **39**, 41, 49, **50**
Borders, 34, 35, 38, 49
Bridal quilts, 12
Broken dishes, **77**
Buttonhole stitch, 23, **23**

Cardboard, 12, 20, 53
Center medallion, **70**
Chair pads, 56
China, 11
Churndash, **68**
Clothing, old, 11, 49
Colonies, American, 11
Color testing, 19
Corn, **99**
Corners, appliqueing, 24; binding, 40, 41
Cotton fabric, 19
Crab, **94**
Crazy block, **12**
Crib quilt, 56; size of, 46
Crosspatch, **59**
Crusades, 11
Curtains, 11, 52
Curves, appliqueing, 24

Daffodil, **84**
Dahlia, **106**
Daisy, large, **105**; small, **106**
Darning, 57
Depression, 12
Diamond pattern quilting, 30, **30**
Double bed, fabric for quilt, 47; quilt size, 46
Dresden plate, **81**
Dry cleaning, 28
Drying, by machine, 28
Ducks-foot-in-the-mud, 11
Dust ruffles, 46

Easing, 14, 41
Embroidery, 43; floss, 23, 57
Europe, 11
Fabric, 12, 46, 49, 57; amount, 47; backing, 28, 29, 38, 39, 44; fading, 57; shortage, 11; type, 19, 57
False mitre, 40, **40**
Filling, polyester, 51, 53
Fish, **103**
Flame, **91**
Frame, quilting, 44
Fraying, prevention of, 27
French knots, 27
Friendship quilts, 12
Frog, **88**

Grain, straight, 40
Grandmother's fan, **75**

Hand of friendship, 11
Hawaiian quilting, 33, **33**, 49
Heart, single, **102**; row, **107**
Hemming stitch, 22
Hibiscus, 92
Homespun, 11
Hourglass, **67**

Ice cream cone, **86**
Indentations, appliqueing, 24, **26**, **27**, 49
Ironing, 14
Italian stuffed quilting, 52, **52**

Jackets, 11
Jacob's Ladder, **72**
Jerusalem, 11
Joining blocks, 37

King-size bed, size of quilt, 46
Knot, 52; quilting, 30

Ladybug, **101**
Lattice strips, 38
Layers, 28
Leaf, **107**
Lemon star, **80**
Lemoyne star, **80**
Lincoln's Platform, 11, 17, **18**, **63**
Locust leaf spray, **109**
Log cabin, 11, **18**
Long Island, 11
Luma-luma quilting, 33, **33**, 49

Matching seams, 14, 56
Missionaries, 49
Modern geometric, 18, **19**
Mushrooms, **89**

Needle, recommended, 30
Nine Patch, 14, **15**, 17, **47**, **65**
Nine sharps needle, 30
Noah's ark, 27

Orange peel, **98**
Outline quilting, 51
Outline stitch, 27
Overall Bill, **60**

Parallel lines, 30, **30**, 32, **32**, 33, **33**, 44, **45**
Pattern pieces, 19, 20, 53
Philadelphia, 11
Pillows, 35, 49, 52, 56
Pinning, 28, 29
Pinwheel, **79**
Points, appliqueing, 24, **24**, **25**
Polyester and cotton fabric, 19
Poplar leaf, **61**
Presentation quilts, 12
Preshrinking fabric, 19
Pressing, 14

Queen-size bed, quilt for, 46
Quilting stitch, 29, **29**, 34
Quilt sizes, 46

Rainbow, **58**
Robbing Peter to pay Paul, **78**
Roman stripe, **80**
Rose of Sharon, 11
Running stitch, 14, 21, **21**, 29, **29**, 30

Sandpaper, 19, 20
Satin stitch, 24, **24**, 27, 49, **50**
Scalloped edges, binding, 40
Seams, 14, 19, 37, 40, 53, 56
Seasons, 27
Sewing machine, 14, 39, 53
Sheets, bed, 44
Shoo-fly, **16**, 17, **48**, **60**, **74**
Silk, 11
Single bed, quilt size, 46
Sizes for quilts, 46
Skirts, 11, 52
Slashing, 26, 27
Snail, **87**
Snake, **87**
Snowflake, **61**, **82**
Star puzzle, 62, **63**, **66**
Stay stitching, 27
Straw, 11
Stuffed, appliqué, 51, **51**; binding, 43
Sunbonnet baby, 90
Superstition, 9, 49
Swirl, 104

Tacking, 25, **25**, 26, **26**, 52
Thimble, 29
Thread, color of, 22, 29, 35; cotton, 14, 30; cotton-covered polyester core, 30; in basting, 29; length, 30; polyester, 14, 49; quilting, 30; single, 30
Tissue paper, 35
Tomahawk, 11, **76**
Tracing, 35, **36**
Trapunto, 52,**52**, 53
Trimming, 14, 37, **37**, 38
Turtle, **85**

Umbrella, **100**
Umbrella girl, **60**
Undergarments, 11

Washing by machine, 28
Washington pavement, 11
Watermelon, **34**, **93**
Weathervane, **62**, **73**
Whip stitch, 22, **22**, 52, **52**
Wool, 11
World's fair, 11
Wreath, **108**

Yarn, 57
Yo-yo, 52